Climate, Environmental Hazards and Migration in Bangladesh

The apocalyptic visions of climate change that are projected in the media often involve extreme weather events, disasters and mass migration of poor people. This book takes a critical look at this notion, drawing on research in Bangladesh, a country located at the heart of debates on climate change and migration.

This book argues that rather than leading to dramatic events, climatic and environmental impacts often cause incremental changes in people's habitats and livelihoods, making them migrate in search of better places and income. With or without climate change, climatic and environmental factors can impoverish people, and drive displacement and migration, especially in the global South. These influences, including disasters, need not necessarily make people move, but instead sometimes trap the poorest and the most vulnerable people in their places exposed to hazards or make them migrate to even riskier places, such as crowded and flood-prone urban slums. This book argues that restrictions placed on people's mobility options could increase their vulnerability and favours proactive migration policies.

This timely contribution explains the climate-hazard-migration nexus in an accessible, engaging language for students of geography, development studies, politics and environmental studies, as well as humanitarian and development practitioners and policymakers.

Max Martin is a geographer focusing on people's responses to climate and environment. He studied at Universities of Kerala, Oxford and Sussex. He has engaged in research at the Institute of Development Studies, UK, and International Centre for Climate Change and Development, Bangladesh; and taught human ecology at University College London. Currently he is a research associate at Sussex, developing a marine weather forecast system called *Radio Monsoon* for artisanal fishers of the Arabian Sea.

Routledge Studies in Hazards, Disaster Risk and Climate Change

Series Editor:
Ilan Kelman,
Reader in Risk, Resilience and Global Health at the Institute for Risk and Disaster Reduction (IRDR) and the Institute for Global Health (IGH), University College London (UCL).

This series provides a forum for original and vibrant research. It offers contributions from each of these communities as well as innovative titles that examine the links between hazards, disasters and climate change, to bring these schools of thought closer together. This series promotes interdisciplinary scholarly work that is empirically and theoretically informed, with titles reflecting the wealth of research being undertaken in these diverse and exciting fields.

For a full list of books in this series please visit www.routledge.com/Routledge-Studies-in-Hazards-Disaster-Risk-and-Climate-Change/book-series/HDC

Published:

Unravelling the Fukushima Disaster
Edited by Mitsuo Yamakawa and Daisaku Yamamoto

Rebuilding Fukushima
Edited by Mitsuo Yamakawa and Daisaku Yamamoto

Climate Hazard Crises in Asian Societies and Environments
Edited by Troy Sternberg

The Institutionalisation of Disaster Risk Reduction

South Africa and Neoliberal Governmentality
Gideon van Riet

Understanding Climate Change through Gender Relations
Edited by Susan Buckingham and Virginie Le Masson

Climate Change and Urban Settlements
Mahendra Sethi

Community Engagement in Post-Disaster Recovery
Edited by Graham Marsh, Iftekhar Ahmed, Martin Mulligan, Jenny Donovan and Steve Barton

Climate, Environmental Hazards and Migration in Bangladesh
Max Martin

Climate, Environmental Hazards and Migration in Bangladesh

Max Martin

Routledge
Taylor & Francis Group
LONDON AND NEW YORK

First published 2018
by Routledge
2 Park Square, Milton Park, Abingdon, Oxon OX14 4RN

and by Routledge
711 Third Avenue, New York, NY 10017

Routledge is an imprint of the Taylor & Francis Group, an informa business

British Library Cataloguing-in-Publication Data
A catalogue record for this book is available from the British Library

Library of Congress Cataloging-in-Publication Data
A catalog record for this book has been requested

ISBN: 978-1-138-23849-7 (hbk)
ISBN: 978-1-315-29745-3 (ebk)

Typeset in Times New Roman
by Apex CoVantage, LLC

MIX
Paper from responsible sources
FSC® C013056

Printed and bound in Great Britain by
TJ International Ltd, Padstow, Cornwall

Contents

List of figures vii
List of tables ix
Acknowledgements xi

1 Introduction 1

2 Bangladesh: a hotspot of climate change 13

3 Apocalypse now? Alarm, skepticism and evidence in the climate-migration debate 23

4 Stay or move? Migration decisions amidst uncertainties 51

5 Climate, environment or poverty? Statistical analysis of factors that drive migration 77

6 Removing roadblocks: policy challenges in making migration an adaptation strategy 103

7 The song of the road: weaving together the migration story 121

Annexes 137
References 205
Index 225

Figures

1.1 The conceptual framework used in the *Foresight (2011)* report on migration 6
1.2 Map of Bangladesh showing the study areas 9
3.1 Model of Private Proactive Adaptation to Climate Change (MPPACC) 37
3.2 Conceptual model of Migration Adaptation to Rainfall Change (MARC) 38
4.1 Map of Bangladesh showing the sub-districts chosen for the qualitative study 55
4.2 Model of migration decision-making 60
5.1 Kernel density graph showing the probability of first migration by age 89
5.2 Kernel density graph showing first migration by age and district 89
5.3 Kernel density graph showing the age of women and men migrating 90
5.4 Kernel density graph showing the age of women migrating, by district 91
5.5 Kernel density graph showing the age of men migrating, by district 91
5.6 Kaplan-Meir survival graph showing migration of women and men 93
5.7 Kaplan-Meir survival graph showing migration with relation to cyclones 93
5.8 Kaplan-Meir survival graph showing migration with relation to droughts 94
5.9 Kaplan-Meir survival graph showing migration with relation to riverbank erosion 94

Tables

3.1	Various projections of climate migrants and refugees	28
4.1	Field study villages and climactic and environmental hazards	54
5.1	Respondents who have migrated outside their district of origin	84
5.2	Respondents who have moved houses	84
5.3	Migrants and non-migrants by their district of origin	84
5.4	Major destinations of the migrants	85
5.5	**Socio-economic variables** (and their assigned values in the logit models)	86
5.6	**Climatic and environmental variables** (and their assigned values in the logit models)	86
5.7	Logit Model A: climatic and environmental variables that influence migration	87
5.8	Logit Model A-m: first migration of men out of the district of origin	96
5.9	Logit Model A-w: first migration of women out of the district of origin	97
5.10	Logit model B: first instance of moving house	98

Acknowledgements

This book is based on my doctoral work at the University of Sussex. Let me first thank the people who made my doctoral work possible. I would first like to thank my supervisors Prof Richard Black and Prof Dominic Kniveton for accepting me as a doctoral candidate, and guiding and supporting me in my research. My masters supervisors at Oxford, Prof Dawn Chatty of the Refugee Studies Centre, and Dr Nicholas Van Hear of the Centre on Migration, Policy and Society, encouraged me to take up doctoral studies; Prof Chatty helped me conceptualise my research. My thanks to Prof Michael Collyer for advising me at Sussex when I started my research, and later as my internal examiner; and Dr Craig Hutton of the University of Southampton for his review and suggestions as my external examiner. Thanks to Prof S Parasuraman and Dr P Vijayakumar of Tata Institute of Social Sciences, and Prof S Japhet of National Law School of India University, for their initial advice.

My research was part of a CDKN-funded project that studied climate change and migration in Bangladesh, coordinated by the Sussex Centre for Migration Research and the Refugee and Migratory Movements Research Unit (RMMRU) at the University of Dhaka. At Dhaka, RMMRU colleagues, led by Prof Tasneem Siddiqui and Prof Chowdhury Abrar, gave me all the support for field research. Later, Sussex Fund and Williamson Memorial Trust gave me grants to help me finish my thesis. Thanks to Mr Terry Cannon of IDS for giving me an opportunity to continue my research in Bangladesh, and Dr Saleemul Huq of the International Centre for Climate Change and Development (ICCCAD) at the Independent University, Bangladesh for hosting me as a visiting researcher. Thanks to the villagers of our field sites for sharing their time and experiences. Thanks are also due to many government officials, NGO representatives and academics of Bangladesh who gave me interviews and advice.

I am grateful to Mr Robert Maidment-Evans, Mrs Marion Maidment-Evans, Mr Ahtsham Rana, Dr Roger Williamson, Dr Antony Lewis, Fr Paul Wilkinson, Ms Margaret Rey; and my friends from back home, Mr Sinoj Mullangath, Ms Darlena David, Dr Unni Krishnan, Mr Premangshu Ray and Mr Rajat Banerjee, for their kindness, friendship and advice. My sincere thanks to my wife Ms Smitha M Martin for her love, patience and all the support she gave me; thanks to our extended families too. Thanks to my son, Master Ashish M Martin for bearing with my frequent absence from home, and changing schools twice to fit my schedule.

As for the book process, my sincere thanks to series editor Dr Ilan Kelman of University College London, who encouraged me to publish this work, Routledge commissioning editor Ms Faye Leerink, editorial assistant Ms Priscilla Corbett and teammates; and the production team led by Ms Kate Fornadel.

1 Introduction

Tracing linkages between climate, environment and migration

Max Martin

1.1 How do you leave a sinking island?

On a monsoon morning, my local guide, a social worker, took me to Gabura, a riverine island in southwestern Bangladesh located close to the Sundarbans, the largest mangrove forest in the world. Our small, low boat chugged across a sprawling river towards what looked like a flat, barren, disc, a deep grey cloud hovering above, pouring down. This place would become a field site for my doctoral research on climate-related migration. Sheets of needle-sharp raindrops pierced through my light waterproofs and backpack. Passing many men and women fishing in even smaller canoes, we reached Gabura in an hour or so. The boatman docked his craft beside a muddy embankment. We climbed over, my guide holding me tight by my arm so that I did not tip over and sink in the water or the ankle-deep mud made of Himalayan silt.

In the village informal education centre, a small building, about 20 local women gathered to talk with us over sweet tea. Their husbands were away, in farms of some other villages, or in some regional town. The men had to migrate to work for four to five months a year. No crop grew in Gabura in 2012. Cyclone Aila of 2009 had hurled a three-metre storm surge at Gabura, breaching the embankment, flooding fields for several months, leaving them saline, barren. The women said they lived in fear. The Bay of Bengal spawns fierce storms once every three years on an average.

Would they move out if the government gave them a better, safer place? Away from the stormy Sundarbans, a forest infested with tigers, crocodiles and river pirates.

"No." The women sounded confident and sure.

"It's our home." "Our ancestors were born, lived and were buried here."[1]

People staying put in such hazard-prone places defined the storyline of my doctoral work. It was part of a CDKN-funded migration research project. I was working with professors and peers at the University of Sussex and the Refugee and Migratory Movements Research Unit at Dhaka University. We studied Gabura and other places exposed to storms, floods and droughts – 14 villages in three districts. The villagers in these places told us that migration is a tried and tested livelihood choice, one among many. They migrated for many reasons, for short and long duration, to near and far places. It helped them escape seasonal poverty, and offset the impacts of floods, cyclones and droughts.

Sometimes a fast-flowing, silt-laden Himalayan river – Ganga, Meghna, Brahmaputra – or a tributary would sink or sandpaper off chunks of a village farm, homestead or even part of a house. In the pre-harvest season of food shortages and hunger, men and boys move out, looking for temporary work anywhere. During a flood or a storm surge, entire families might find refuge on an elevated road or an embankment. Farm losses, food shortages, water scarcity and soil salinity could undermine livelihoods, and people would migrate to earn better. Movement is a way of life in many parts of Bangladesh.

People who move out often come back when the floods subside or the spell of drought is broken. They gather just enough money to repair the damaged house, or start farming again on a rice field rendered salty and barren by a storm surge. Sometimes young women and teenage girls go out to work in the garment factories of Dhaka or Chittagong, earning enough to support the family or to save for their future life.

We found that most of the time, migration is a way to earn better. The latest (2011) census shows people moving out from the relatively less developed southwestern coastal belt, even from Khulna city, the third biggest in the country. Poverty in villages and rapid growth in cities provide the push and pull for migration. Experiences of climate and the environment appear to work in the background – often not making a direct, linear link, as it is often made out to be.

However, the predominant narrative, especially in popular media, is all about year-to-year climate variability and longer-term climate change making people move out permanently – to faraway places, even to western countries. The story we find, however, is that migration is a complex phenomenon, driven by overlapping influences. It is often local and short-term.

In a future that is wetter, warmer and uncertain, these patterns could dramatically change. Even then speculating that people living in places like Gabura on the margins of the map will catch the next flight out of the country is a tabloid fantasy. The notion of masses of "climate refugees" is often a media myth. It is not to suggest that migrants do not cross state borders – they do, but largely migration is local, even temporary, as we found in Bangladesh.

At the same time, disasters could also wipe out resources needed for migration, forcing people to stay back, trapped. A key concern addressed in this book is immobility: people being unable or unwilling to move out of their homes in the face of life-threatening hazards – cyclonic storm surges, recurrent floods and so on. Often people lack resources to migrate. Moving out and settling in a new place costs money. Sometimes, people do not realise or refuse to believe that their place could be exposed to such high levels of vulnerability that it may be almost impossible to escape the next disaster. Some people, such as estuarine fisherfolk or those who collect honey and other minor forest produces from the mangroves, may be tied to a particular place for their livelihoods. Still some others are attached to their place – just would not leave their place of birth, where their folks and friends live.

1.2 The storyline

Bangladesh is a country at the heart of debates about climate change and migration. This book probes to what extent climate- and environment-related hazards influence decisions of villagers in Bangladesh to stay or move out of their place. It considers their experiences of hazards such as cyclones, droughts and floods as proxies of what might happen in the future as a result of changing climate and probes how they respond when their livelihoods are affected by these stresses and shocks.

The qualitative analysis in this book shows that villagers from three hazard-prone districts of Bangladesh – Nawabganj, Munshiganj and Satkhira – often migrate for better livelihoods. However, they usually do not associate their movement to the hazards. At the same time, the quantitative analysis done for this research shows that experiences of drought and cyclone positively influence migration outside the district. Though riverbank erosion and flood negatively influence long-distance migration, people affected by erosion tend to move locally. Logit models suggest that though migration is largely driven by poverty and income needs, the poorest, especially those without any assets, are often unable to migrate outside the district. Meanwhile social networks and education contribute to migration.

Whether people state it or not, migration can be a strategy that helps them offset losses and prepare better for future stresses and shocks. However, whether such migration leads to adaptation to climate change depends on the policy environment in the country. A textual analysis of policy documents, however, shows that though urban migration is inevitable for Bangladesh's economic growth, its role as a climate change adaptation strategy is often not acknowledged. The book argues that policies need to be more proactive so that migration does not become something that makes people more vulnerable or unable to move out, trapped in places exposed to climate- and environment-related hazards.

1.3 Linking climate, environment and migration

> "Climate change is projected to increase the displacement of people throughout this century. The risk of displacement increases when populations who lack the resources to migrate experience higher exposure to extreme weather events, in both rural and urban areas, particularly in low-income developing countries. Changes in migration patterns can be responses to both extreme weather events and longer-term climate variability and change, and migration can also be an effective adaptation strategy."
>
> (IPCC 2014: 73)

The above statement from the latest synthesis report of the Intergovernmental Panel on Climate Change (IPCC) sums up the context of this book. The book explores how people in villages of Bangladesh make decisions to stay or move out of their place when faced with changes in their climate and environment. While climate comprises weather conditions prevailing in an area over a long period, environment encompasses the surroundings and conditions in which a community lives and earns a livelihood. Climate and environment are interconnected and closely interact with human systems. The literature suggests that globally human-induced impacts on the composition of the atmosphere, climate, water and land resources and biodiversity are occurring so rapidly that natural systems are often unable to adapt to these changes (UNEP 2003). These changes – often broadly termed as global environmental change – can affect livelihoods, food security, habitats, health and well-being (UNEP 2003). Climate change is a major part of this phenomenon. Bangladesh is a theatre where the drama of climate change unfolds in an exaggerated manner, in superlative terms, as the next chapter narrates.

To continue living as a viable social group in the face of such dramatic changes in their local climate and environment, people may have to adjust, adapt and rebuild (Oliver-Smith 2009) or move out to a different place faraway or nearby for a long or short period (McLeman and Smit 2006; Tacoli 2009; Barnett and Webber 2010). Global environmental change enhances disaster and impoverishment risks, and it is expected to almost certainly alter human migration patterns in the coming century (Warner et al. 2010). However, large-scale migration is not necessarily the consequence of environmental change; there are multiple drivers behind human movements.

A new thinking that is gaining currency in this field is that while migration continues regardless of environmental change, the latter could still influence current and future migration patterns through a range of complex interactions, as an influential UK Science Office report argues (Foresight 2011). Titled *Migration and global environmental change: future challenges and opportunities*, the report argues that even when climatic and environmental factors drive migration, a large part of such migration will be in the Global South, within countries or to nearby countries, including to areas of environmental risk (Foresight 2011). In such scenarios, preventing or restricting migration could be risky in the sense that immobility could further impoverish people or lead to displacement and irregular migration in future (Foresight 2011; IPCC 2014). This is one key concern that this book addresses.

As such, there is considerable local and global concern about climate change and climate variability and their impacts on mobility. Different countries across the world are considered vulnerable to natural hazards and climate variability. "Climate change refers to any change in climate over time, whether due to natural variability or as a result of human activity. Climate variability refers to variations in the mean state and other statistics (such as standard deviations, statistics of extremes, etc.) of the climate on all temporal and spatial scales beyond that of individual weather events. Variability may be due to natural internal processes within the climate system (internal variability), or variations in natural or anthropogenic external forcing (external variability)" (IPCC 2007a). Recent research has highlighted different mobility outcomes and their implications under various climate change scenarios (IPCC 2014). At the same time, climate variability also needs attention, as often the potential damage from existing climate variability could be many times more than that of projected impact of climate change in the case of countries such as Bangladesh (Adams et al. 2011**a**).

1.4 Focus on a climate change hotspot

Perhaps nowhere are links between climatic and environmental factors and mobility pronounced more than in Bangladesh, a country that sits at the heart of debates about climate change impacts. The country is considered vulnerable to extreme weather events and weather uncertainties due to its geographic, socio-economic and demographic features. It has been facing gradual onset climate stresses and sudden shocks, including water shortage, cyclone, floods and coastal/riverbank erosion (Adams et al. 2011a). It is not only climate extremes that matter; even non-extreme events in a background of social vulnerabilities and exposure to risks threaten people's lives and livelihoods (IPCC 2012). Overall, the country is considered a climate change hotspot (Huq 2001; Huq and Ayers 2008), that means a place that is particularly vulnerable to the impact of climate change now or in the future with considerable risk to human security (de Sherbinin 2014).

During 1994–2013, Bangladesh was ranked among six countries most affected by extreme weather events (Kreft at al. 2015). Hazards such as flood, cyclone and drought often made worse by climate change (Huq 2001, Huq and Ayers 2008) also influence existing migration patterns in Bangladesh (2007; Gray and Mueller 2012; Black et al. 2013; Penning-Rowsell et al. 2013). However, a large share of migration is driven by high levels of rural poverty in the context of rapid urban growth (Muzzini and Aparicio 2013; Marshall and Rahman 2013) and high population density (BBS 2011). Such dimensions of the environment-climate-migration nexus require a closer look.

However, these multiple dimensions were overlooked in most of the earlier literature that links climate change with migration. One reason for such a narrow view was that apocalyptic projections of climate change amidst scarcity of natural resources in a growing world have often led researchers to propose staggering figures of migration, directly driven by climatic and environmental factors (Tickell 1989; Homer-Dixon and Percival 1996; Myers 2002 for instance). Later research, however, has questioned the empirical evidence base behind such figures and debated the approach that produced them (McGregor 1993; Black 2001; Castles 2002; Gemenne 2011; Jakobeit and Methmann 2012). A more nuanced view sees migration as a survival strategy (McGregor 1993) with a broad spectrum of causes and consequences (Black 2001), influenced by a set of socio-economic and political factors. This book acknowledges such a complex relationship, specifically the influence of climate- and environment-related factors in migration decision-making.

Though this book acknowledges the context of climate change in Bangladesh, it desists from using the terms 'climate change migration' or 'climate-induced migration' and coins a more neutral and inclusive phrase, 'climate- and environment-related migration.' The reasons for avoiding a strong association of migration to climate change have been explained in detail in chapter 3 that comprises the literature review. The phrase climate- and environment-related migration denotes migration in the context of climatic and environmental stresses, shocks and changes, possibly influenced by them to some degree positively or negatively, directly or indirectly, but not necessarily induced by these factors.

1.5 Concepts, aims and objectives

1.5.1 Climate- and environment-related migration

As this book looks at climate- and environment-related migration in a context of a country facing the impacts of climate change, it adopts the conceptual approach outlined in the Foresight (2011) report. According to this approach, environmental change affects drivers of migration spread across social, political, economic, environmental and demographic spheres (as shown in figure 1.1) in different degrees. These drivers may or may not lead to migration, depending on how a set of enabling and limiting factors influence the decision-making regarding movements. The merit of this approach in relation to the book is that it puts together different external and internal influences of migration in a comprehensive framework. Therefore, it aids not only testing the sensitivity of migration to climatic and environmental factors, but also tests whether other elements influence migration decision-making.

A weakness of the Foresight framework (figure 1.1), however, is that it does not take into account personal experiences of hazards and perceptions of risk that mediate migration decisions (as analysed in Kniveton et al. 2011, for instance). While considering the sensitivity of existing drivers of migration to climate and intervening factors, this book considers how experiences and perceptions influence people's decisions to move out of their place or stay put there. It acknowledges that migration decisions depend on a series of intervening factors and personal and household characteristics (Black et al. 2011a) that work at different levels – from personal and household levels to regional and structural (Schmidt-Verkerk 2011). In a context of changes and variability in climate and environment, especially in Bangladesh, this research looks closely at the decision-making process on the right side of the diagram; and how these decisions were influenced by experiences of hazards and their impacts shown on the left corner of the pentagon. These two points have been taken into account to understand how people's

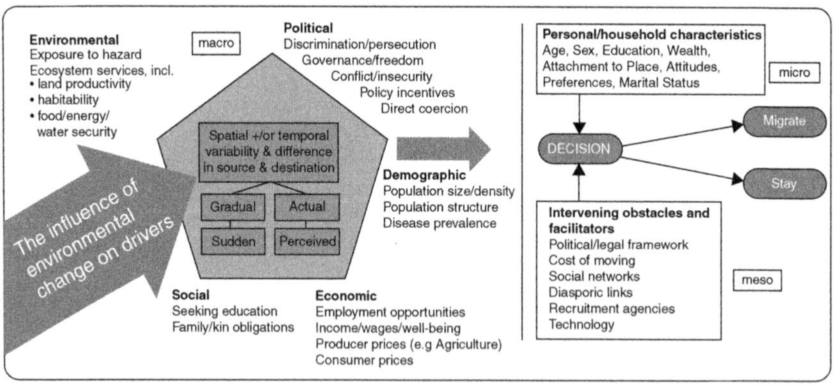

Figure 1.1 The conceptual framework used in the ***Foresight (2011)*** report on migration

Environmental change influences multiple drivers of migration decision-making. It is as likely to restrict migration as it is to cause migration (Foresight 2011)

experience of hazards and the perceptions of the risk they pose influence migration decision-making process.

On another plane, the book counters the assumption of some of the earlier research on migration associated with climate change. These studies assumed that stresses and shocks translate directly into migration (for instance Tickell 1989; Homer-Dixon and Percival 1996; Myers 2002). This book probes how movement influenced by climatic and environmental reasons happens in a general context of mobility, and to what extent income and livelihood imperatives also determine migration. The literature shows that people migrate for a variety of reasons that include better income, reduced risks, offsetting market losses and more work opportunities (Stark and Levhari 1982; Stark 1984; Stark and Bloom 1985; Massey et al. 1993). Still, exposure to environmental hazards and disasters may recast a place with a negative image and influence people's relocation decisions (Hunter 2005). Still these movements could mean forced migration in the immediate aftermath of a disaster, or because of a need for better livelihoods and income necessitated by challenges that their local climate and environment pose.

1.5.2 Migration decision-making

This research follows up on recent enquiries into this process of migration decision-making. The literature shows that migration decisions are mediated primarily through individual agency though it is part of a household-level decision-making process supported by social networks. As Kniveton et al. (2011: S36) put it succinctly, behind each individual decision is a "unique combination of experiences, biases, assets and perceptions." These unique experiences explain the heterogeneity of migration decisions. Theoretically speaking, an individual migrant's intentions and behaviour are shaped by his or her attitude and a set of beliefs and perceptions (Ajzen 1991) as well as a thought process that includes risk appraisal and adaptation appraisal (Grothmann and Patt 2005).

1.5.3 Aims and objectives

The book traces the influence of local climate- and environment-related hazards on migration of villagers in Bangladesh. As a corollary, it explores questions regarding various climate- and environment-related stresses and shocks; changes and uncertainties in climate and environment and the risks involved; people's acknowledgement of these factors; statistical relationship of these factors with long-term migratory movements; and the policy environment that deals with climate-related migration.

Overall, the book aims to understand the influence of climate- and environment-related hazards on migration from villages in Bangladesh. Its objectives include answering a set of corollary questions regarding different dimensions of climate- and environment-related migration. They include:

1 Various climatic and environmental stresses and shocks in rural Bangladesh.
2 People's experiences of changes and uncertainties in climate and environment; and their perceptions of risk.

3 People's acknowledgement of the influence of hazard experiences and risk concerns in their migration decisions.
4 Statistical relationship of climate- and environment-related hazards with long-term migratory movements.
5 Policy-level acknowledgement of the role of migration as a climate change adaptation.

These research questions have been discussed in detail in chapter 4.

1.6 Data gathering

The data for the research largely came from a climate-related migration project of the Sussex Centre for Migration Research at the University of Sussex and the Refugee and Migratory Movements Research Unit (RMMRU) at the University of Dhaka, supported by CDKN (CDKN 2011). The author has worked as the graduate research assistant in the project, contributing to research design, coordination, field interviews, training of field interviewers and data analysis, and as the lead author of a set of publications.

The research has covered areas perceived to be climate change hotspots in recent literature spread out in three districts. They are flood plains and riverine villages in Srinagar, Sirajdikhan and Lohajong sub-districts of the central Munshiganj district; drought-prone villages in Nachole, Shibganj and Nawabganj sub-districts of Nawabganj in the northwest; and cyclone and flood-affected places in Shyamnagar sub-district of Satkhira in the southwestern coastal zone (Rahman et al. 2007; Walsham 2010; World Bank 2010).

As for data collection, secondary data has been used to identify key climate- and environment-related characteristics as well as broad migration patterns of Bangladesh. Primary data includes village-level focus groups, interviews and qualitative and quantitative surveys, policy documents as well as rainfall data from meteorological observatories and flood and cyclone data compiled from government reports. The qualitative field data considers how individuals and households make migration decisions based on a number of factors that include livelihood challenges, economic needs, experience of climate-and environment-related hazards and the perceptions of risks they pose. It also looks at the socio-cognitive variables that influence migration decisions. Questions focus largely on perception and assessment of environmental risks, and barriers and facilitating factors of migration. It explores the decision-making process and to what extent migration is an effective adaptation strategy. Meanwhile policy documents – in the fields of migration, climate change, development and disaster risk reduction – provide the basis for a textual analysis of government attitudes to climate- and environment-related migration, and identification of enabling and disabling factors for prospective migrants. The quantitative data covers questions on migration, employment, assets base, demographics, community-level and environmental data. This analysis considers the combined effect of different socio-economic as well as climatic and environmental determinants of migration in response to climatic stresses and shocks. The analysis

Figure 1.2 Map of Bangladesh showing the study areas

also reveals the characteristics of those most likely to migrate in the face of climate-related threats and environmental change.

1.7 Contribution

The contribution of this book lies in three areas of research into climate- and environment-related migration. First, it gathers empirical evidence for linking migration decisions with climatic and environmental factors using qualitative and quantitative methods. Second, it looks closely at the migration decision-making process itself. Third, it considers migration as a climate change adaptation strategy and looks at the policy implications of such a viewpoint.

Theoretically, the book aims to frame environmental change as a phenomenon that influences existing multiple drivers of migration (Black et al. 2011a, Foresight 2011). It traces the roots of such a framing back to neoclassical migration theories (for instance, Todaro 1969). However, it identifies their limitations in the context of climate-related migration. While the book acknowledges the livelihood and economic dimensions of migration, it places the influence of climatic and environmental factors upfront, unlike in studies that take a neoclassical approach. Second, it argues that while migration is driven by poverty and income needs, the poorest, especially those without assets, are often unable to make long distance migratory movements. Third, it goes beyond the framing of behavioural factors of migration (for instance, Wolpert 1965; De Jong and Fawcett 1981) by incorporating elements of cognitive enquiry (Grothmann and Patt 2005; Kuruppu 2009; Kniveton et al. 2011, 2012; Reckien et al. 2013).

In terms of practical relevance, the book considers different forms of human mobility in a spectrum ranging from displacement to planned urban migration for economic gain. It considers how people view migration as a strategy for more remunerative livelihoods and as a way to offset losses suffered in disasters, and to be prepared for future stresses and shocks. The argument here is that as people view it as a positive step, such migration could be seen as contributing to adaptation in the context of climate change. Therefore, this book argues for a policy realignment that acknowledges the adaptive role of migration. While displacement is often inevitable, policies could help people avoid it by aiding alternatives such as local adaptation, migration and, to a limited extent, planned relocation.

Rather than proving or disproving how climatic and environmental drivers cause migration, this book attempts to assess the sensitivity of existing drivers and intervening factors affecting migration to climate change. It focuses on existing migration trends in Bangladesh and considers their climate- and environment-related influences. To a limited extent, it uses an integrated assessment approach looking at climate impacts and vulnerability assessment and policy analysis (Black et al. 2011b). It represents interactions across different spatial and temporal scales, climatic processes and events such as rainfall variability, droughts and storms and people's livelihood activities. It includes statistical models, qualitative field research and text analysis of policies. While a mix of qualitative and quantitative approaches helps to triangulate the findings, the book offers new insights into the climate-environment-migration nexus.

1.8 Limits

This book does not attempt to study long-term climatological patterns and their possible impacts on people's mobility patterns. Rather, the focus is on climate- and environment-related hazards over the recent past, based on retrospective data that goes back around two decades on average. Second, its inferences are largely based on surveys using both quantitative and qualitative methods. It mainly depends on how individuals report changes in their environment and climate and how they affect their livelihoods and influence their mobility. Information about the household dynamics of decision-making has been elicited from these individual interviews – a more rigorous approach involving more family members in follow up discussions would have given a more detailed picture of household-level decision-making.

Further, the use of instrument data has been limited to rainfall measurement, as well as country-level data on cyclone and floods. A more spatially disaggregated data based on instrument observation would have contributed to a more rigorous analysis – but that was beyond the scope of the research project. No attempt has been made to discern long-term trends that could be proven as climate change – we have clustered year-to-year variability in hazards and longer-term trends to understand their impact on migration decisions. These observations may not be considered as proof for climate change impact or otherwise. Last, but not least, more in-depth field enquiries on these influences and linkages would have helped in triangulating the findings. However, follow up field visits had to be ruled out in view of political troubles and violence in Bangladesh in 2013, during the last phase of the project period (FCO 2014).

1.9 Structure

After this introduction, chapter 2 is a case study of Bangladesh that describes the country's hazard exposure, migration patterns and drivers. Chapter 3 provides a literature review to place the book in the context of research broadly in the field of climate- and environment-related migration. This chapter first looks at the evolution of the concept of the environment-climate-migration nexus, reviews migration theories and connects the book to the literature in behavioural sciences, cognitive analysis and adaptation; then it reviews empirical studies in the field.

The following three empirical chapters (4–6) respectively adopt a qualitative, quantitative and policy analytical approach to answer the core research questions outlined above. Chapter 4 uses a qualitative analysis approach to understand how experience of changes and uncertainties and risk perceptions influence people's migration decisions. The following chapter 5 uses quantitative analysis to probe how experience of climate- and environment-related hazards relate with long-term migration. Chapter 6 takes a close look at the climate change, migration, development and disaster risk reduction policies of Bangladesh, and examines to what extent they acknowledge migration as a climate change adaptation measure. Finally, chapter 7 synthesises the findings, and concludes the book.

Note

1 Edited version of 'How do you leave a sinking island?', a blog post by the author that appeared in Thomson Reuters Foundation blog.

2 Bangladesh

A hotspot of climate change

This chapter is an introduction to Bangladesh and its geographic features that make it vulnerable to the impacts of climate change and associated extreme and uncertain weather events. Located at the confluence of three major Himalayan rivers, much of this densely populated country's low-lying plains, coasts and riverine islands are prone to frequent flooding. Its southern coast is exposed to devastating cyclones from the Bay of Bengal. Climate models predict a wetter future and more intense cyclones. Together these impacts render many places too risky to live. They also destroy farming and fishing livelihoods of poor people in a least developed country, who are among those least responsible for climate change. The chapter stresses that it is not just the dramatic changes associated with climate change that make people vulnerable, but the existing hazards patterns in a context of their impoverishment and marginality. It addresses the wider, international relevance of the book, as Bangladesh can be considered a case study where climate change impacts are very much felt on the lives and livelihoods of people.

2.1 Introduction

Bangladesh (figure 2.1) is a country located at the confluence of three major Himalayan rivers, where they form the world's largest delta, often described as a climate change 'hotspot' (Huq 2001; Huq and Ayers 2008). The country's low altitude, presence of a network of three great Himalayan rivers, and exposure to tropical cyclones in the Bay of Bengal make it particularly exposed to climatic stresses and shocks. Recent literature describes it as highly vulnerable to stresses and shocks associated with climate variability and change (Adams et al. 2011a; Poncelet et al. 2007). This vulnerability in the context of 31.4 per cent of the people living under the national poverty line in 2010 (World Bank 2015) and rapid city-based growth leads to large-scale rural-urban migration in Bangladesh (World Bank 2012; Muzzini and Aparicio 2013). Such migration is expected to increase in the context of climate change (Foresight 2011).

The geographic and demographic features of Bangladesh offer challenges in terms of finding linkages between climate, environment and migration, and provide a setting for an interdisciplinary study. The first part of this chapter briefly describes hazards to which Bangladesh is exposed, including climatic stresses and shocks, extreme weather events and environmental change. The second part looks at patterns of migration in Bangladesh. The third part looks more closely at the social, political, environmental, economic and demographic drivers of migration using the Foresight (2011) framework as described in figure 1.1 in chapter 1.

2.2 Hazards

Bangladesh is exposed to a wide range of climatic and environmental hazards, including floods, riverbank erosion, cyclones, food shortages, freshwater scarcity or soil salinity (Poncelet et al. 2010). A large part of Bangladesh consists of floodplains of major rivers, including the Ganges, Brahmaputra and Meghna. Flowing from the 'Water Tower of Asia,' the Ganges-Brahmaputra-Meghna system together has one of the largest catchments in the world, draining an area of about 1.7 million square km (FAO 2011a). It is spread across the great Gangetic plain of northern India and southern Nepal and Nepal Himalaya, the Brahmaputra basin extending northward through Assam and Bhutan and then westward between the Himalaya and the Tibetan Plateau (FAO 2011b). Rainfall variations in this area can change the river flow patterns of Bangladesh.

In its 2900 km course down from the Angsi glacier 5,210 metres high, the Brahmaputra, for instance, flows through the world's deepest valley, around the largest riverine island and some of the wettest places in the world before joining the Ganges and then Meghna, weaving a mesh through their delta before touching the Bay of Bengal. The Brahmaputra is 10 km wide at some places. The Ganges originates in the western Himalayas, and its 2,525 km course runs through the Gangetic Plain of North India and then into Bangladesh, where the main branch is known as the Padma. The Jamuna River, the largest distributary of the Brahmaputra, joins the Ganga, and further downstream, the Padma joins the Meghna. During the summer monsoon – when Bangladesh gets 80 per cent of its yearly precipitation – these rivers have a combined peak flow of 180,000 m_3/sec, the second highest in the world (Adams et al. 2011a).

The floodplains of these rivers sustain the livelihoods of millions of farmers, but they are also highly prone to both inland flooding and riverbank erosion. In catastrophic flood years, such as 1987, 1988, 1998, 2004 and 2007, about 39, 61, 68, 38 and 42 per cent of the area was inundated (Adams et al. 2011b). In many cases, inundation continued for nearly three months (CEGIS 2002). The economic cost of these events on Bangladesh is huge, with estimated losses and damages from the 1998 event alone crossing US $2 billion or 4.8 per cent of GDP.

The Himalayan rivers are known not only for their flow rate and floods, but also for their high velocity and the large amounts of silt they carry, leading to a combination of accretion and erosion that leads to new formation of new land areas and disappearance of existing land. Meghna Estuary has lost about 86,000 ha of land during the 1973–2000 period (Adams et al. 2011a). It has been estimated that between 2,000 to 3,000 kilometres of riverbank experience major erosion annually (Islam and Islam 1985). While past climate shocks have exerted a heavy toll on lives and livelihoods in Bangladesh (Narayan et al. 2000), future climate shocks and stresses are predicted to result in increased flooding, riverbank erosion and salinisation of water resources (Adams et al. 2011a).

Global warming leads to more rain and snowmelt in their upper catchment, or the area drained by these great Himalayan rivers. Extreme rain downstream makes flood impact even worse. The new projection of floods shows more run-off into five great Himalayan rivers – Indus (Sindhu), Ganges (Ganga), Brahmaputra, Salween and Mekong (Lutz et al. 2014). Though there are differences between the river basins and even between tributaries within each basin, overall projection shows an increase in run-off until 2050. That is due to an increase in the precipitation in the Upper Salween, Ganges, Brahmaputra and Mekong basins and more snowmelt in the upper Indus basin (Lutz et al. 2014).

Earlier research has noticed that extreme rain events are also on the rise in the Himalayas, and elsewhere in the subcontinent. A study of hourly precipitation data from 1980 to 2002 across India showed that the north-western Himalaya and northern parts of the Indo-Gangetic basin along the foothills of the Himalaya showed greater frequency of extreme rainfall (Roy 2009). Earlier studies have linked global sea surface temperature patterns, rainfall over India and the stream flow of Ganges and Brahmaputra. For instance, Webster (2010) notes that seasonal discharge of the Ganges is connected with the phase of the El Niño–Southern Oscillation (ENSO), a disruption of the ocean-atmosphere system in the tropical Pacific that influences weather and climate and affects the total Indian rainfall. Summer flow in the Brahmaputra shows a relationship with sea surface temperature in the Indian Ocean as well as the Northwest Pacific and snow depth during the previous spring (Webster et al.2010). In summary, the river flow patterns in Bangladesh could be affected by a large set of factors that are climate sensitive, and there is a trend of a rise in extreme rainfall events in the catchment areas of the great Himalayan rivers.

Cyclones are another major concern. In the coastal villages of Bangladesh, cyclones pose a major threat because of their geographical reach and lingering after-effects. Cyclone Sidr of 2007, for instance, caused 4,234 deaths and affected livelihoods of 8.9 million people (EM-DAT 2015a). Two years later Cyclone Aila caused 190 deaths and affected 3.9 million people (EM-DAT 2015a). Aila's wind speeds ranged from 74 to 120 kmph (NASA 2009). These casualty figures were significantly less than some of the earlier cyclones that hit the country, as disaster risk reduction programmes, including early warning systems, appear to have saved many lives. However, more than 400,000 people were displaced in each of these events that caused prolonged damage to agriculture, fisheries, forestry, health and water supply, forcing people to move out (Roy 2011; OCHA 2012). Current estimates suggest that cyclonic storm surges might cover an additional 15 per cent of the coastal area in the next 50 years (Adams et al. 2011a).

Research in Bangladesh suggests that the impacts of cyclones and storm surges on infrastructure and habitats depend not only on the hazard exposure and intensity, but also demographic factors, along with socio-economic and cultural factors (Paul and Routray 2011). For instance, while very intense tropical cyclones cause heavy damage in low-lying coastal areas in general, irrespective of socio-economic levels, at lower intensities the poorest still suffer heavy losses, whereas the rich are less affected (Peduzzi et al. 2012). Moreover, though the frequency of tropical cyclones may be reducing, projected increases in their intensity could have serious consequences (Peduzzi et al. 2012).

3.3 Migration patterns

To understand climate- and environment-related migration, it is important to have an understanding of the general migration patterns in a country. In Bangladesh, people have traditionally migrated for a variety of reasons. There is a historicity to different forms of movement in and from the country. From pre-colonial times, migrants from the west (now part of India) were attracted by the fertile but wet land of the east; and people have moved in the other direction for trade and to work in plantation and other services. As Gardner (2009: 233) notes: "These constant, cross-cutting migrations are both a result of the region's turbulent history, and its turbulent environment, in which floods and cyclones mean that 'belonging' can never be guaranteed." Such a wide practice of migration is often seen as a way to offset seasonal deprivation (Siddiqui 2009), recover from disasters and other natural hazards (Hunter 2005) and for better living standards and social status (Gardner 2009). Bangladesh has seen four broad types of migration covering these motives – internal movements from villages to other villages or more frequently to big cities; cross-border migration to India; short-term international migration, mostly to the Gulf countries on labour contracts; and longer-term or permanent settlement in western countries.

Among these movements, rural-to urban migration, largely driven by rapid city-based industrial development, is the most prevalent form of movement (Muzzini and Aparicio 2013). Urban growth has been uneven in Bangladesh, with the Dhaka metropolitan area with an estimated 15 million population by far the largest, followed by Chittagong with around 5 million people. These two big cities along with Khulna, Rajshahi, Barisal, Sylhet, Comilla and Rangpur account for about 36 million out of Bangladesh's 160 million population, largely on account of a rising share of industrial production in Bangladesh's GDP (Muzzini and Aparicio 2013). This urban boom is largely driven by unequal growth patterns in the country. Recent analyses show that coastal areas have shown slow growth, with the districts of Khulna and Barisal recording much lower rates of growth compared with the national average (Marshall and Rahman 2013).

The 2011 census (BBS 2012) notes that Bangladesh has an internal migration rate – defined as lifetime migration outside each district per thousand people – of 9.7, with rural-to-urban movements comprising 4.3, rural-to-rural 4.2, urban-to-urban 0.86 and urban-to-rural 0.36. The international migration rate is 3.46 (BBS 2012). Migration – of all of these types – can form an important source of income. In north-western Bangladesh, a 1,600-household survey carried out as part of the Livelihood Monitoring Project found that 19 per cent of households migrated in the lean farming season to supplement their income (Care Bangladesh and DFID 2002). Major rural-to-rural migration takes place during sowing and harvest seasons.

Significant flows of migration have occurred in Bangladesh in recent decades in the context of changing opportunities for employment and income generation. The share of agriculture in GDP fell from 32 per cent to 19 per cent during 1980 to 2010, and that of industry grew from 21 per cent to 28 per cent (Muzzini and Aparicio 2013). That has meant fast growth of cities. In the past two decades, overall population in the country grew by 29 per cent, showing a 24 per cent increase in rural areas and 49 per cent increase in urban areas, with an urbanisation rate of 3 per cent per annum, one of the highest in the world. The corresponding annual rates of increase were 1.3 per cent for Bangladesh, with rural 1.1 per cent and urban 2.0 per cent. An uneven urban growth has led to large-scale movement to Dhaka and Chittagong, as the census figures show (BBS 2011). In the cities, many migrants find jobs as rickshaw-pullers and in informal sectors such as brick kilns and construction or are self-employed in urban and peri-urban areas. In Dhaka, rickshaw-pullers alone numbered about 500,000 in 2005, constituting a total of 2 million people, including their dependents and others closely associated with them (Kreibich 2012).

During 2001–2011, the peri-urban hinterlands of Dhaka and Chittagong have shown major growth, possibly due to saturation of the urban core. Meanwhile coastal areas have shown slow growth, with the districts of Khulna and Barisal recording much lower rates of growth compared with the national average (Marshall and Rahman 2013). Khulna district has shown population decline during 2001–11, with a notable decline in urban population (BBS 2011), suggesting that people are migrating out of both the rural hinterland and the city itself (Marshall and Rahman 2013). However, Bangladesh Bureau of Statistics, an organisation under the Planning Commission that conducts the Census, has explained that the decline in urban population is due to a changed definition of city areas:

> "In earlier censuses, the urban area included city corporations, municipalities, upazila headquarters, growth centers, cantonment and urban agglomerations adjacent to large cities, i.e., city corporations termed as Statistical Metropolitan Area (SMA). In 2011, the concept of SMA, growth center and some other urban areas was abandoned. . . . Due to the definitional change, the percentage of urban population has decreased and come down to 23.30% in 2011 as against 23.53% in 2001. If the same areas would have existed in 2011, the percentage of urban population might have been 28.00% in 2011."
> (BBS 2014)

International migration is another dimension of the migration scenario. The migrants' top destinations include India, Saudi Arabia, UAE, Qatar, Oman, Bahrain, Kuwait, Libya, Iraq, Singapore and Malaysia. (Miah et al. 2014). International remittances have been growing even during the global slowdown after 2008 (Siddiqui 2009). Around 8.7 million work abroad, and they sent US $14.46 billion, or equivalent to 11.14 per cent of the GDP, as remittances in the financial year 2012–13 (MOF 2014). During 2013–2014, the remittances were lesser, nearly US $11.985 billion, or roughly 8 per cent of the country's total GDP of US $150 billion (Bangladesh Bank 2015; World Bank 2015). Remittances have helped reduce rural poverty, and there is an argument that more emigration could improve the country's economy (Moses 2009). International movement also includes crossing the border to India. Historically there have been migratory and refugee movements between eastern and western parts of Bengal. The 2001 Indian national census counted 3 million Bangladeshi migrants, though some estimates suggest higher figures. There has been continued migration to India from the Khulna region, partly due to the effects of the Farakka Barrage on the Indian side that diverts part of the Ganges flow, it has been argued (Swain 1996).

2.3 Drivers of migration

The large-scale movement to cities of Bangladesh is driven by different motivations. Studies suggest that migrants include people trying to escape seasonal deprivation (Chowdhury et al. 2009), recover from the impacts of natural hazards (Hunter 2005; Penning-Rowsell et al. 2013), offset the projected effects of climate change (Tickell 1989; Homer-Dixon and Percival 1996;) and gain better social and economic status (Gardner 2009). Recent studies have examined complexities involved in determining the sensitivity of climatic factors in migration. When village-based farming and fishing activities are sensitive to climatic stresses and shocks (Rahman et al. 2007) as well as demographic and economic pressures (Black et al. 2011), people resort to short- or long-term migration.

The motives and patterns of migration could vary – it could be pre-harvest, seasonal migration or short-distance movement following floods, riverbank erosion, cyclones, food shortages, freshwater scarcity or soil salinity (Poncelet et al. 2010). Even though people do not cite environmental degradation or climatic stresses as a major driver of migration per se, they note how these phenomena have affected their livelihoods, denoting intertwining linkages with other migration drivers (Poncelet et al. 2010). Gray and Mueller (2012) note that while household shocks reduced resources for migration, sub-district-level shocks had a wider effect with little impact on household finances, thus driving more migration (Gray and Mueller 2012). This study makes a distinction between household and community-level migration decisions and movements within a person's district of origin and long-distance destinations (Findlay 2012). Combined with demographic pressures, it could increase the number of people exposed to risk and contribute to more migration.

By working in different locations for short or long periods before returning, migrants try to earn more and save enough to help themselves and their families back home. Yet at the same time, short-term and circular internal migration can be seen as an adaptation strategy for households in districts affected by droughts, cyclones and floods as climatic stresses and shocks undermine villagers' livelihoods (Poncelet 2007; Findlay and Geddes 2011; Gray and Mueller 2012). A proportion of these migrants move short distances to other villages or nearby towns. Others migrate to major metropolitan cities. Though economic push and pull factors are significant in driving migration, changes in livelihood patterns are also influenced by work opportunities as well as vulnerabilities (Entwisle et al. 2005), and migration is a reasoned response embedded in social lives and livelihoods of people (Gardner 2009).

The narrative of climate- and environment-related migration in Bangladesh is rather nuanced. On the one hand, as noted in section 3.2, climate models forecast more rains and an increase in river run-off in Bangladesh, flooding, riverbank erosion and salinisation of water and soil could increase, and is expected by some to lead to more migration (Laczko and Aghazarm 2009;). At the same time, people in climate-sensitive areas increasingly adopt secondary livelihoods that are not dependent directly on natural resources (Ahmad 2012), leading to an increasing trend of urban migration (Afsar 2003; Muzzini and Aparicio 2013; Planning Commission 2011). Often climatic and environmental stresses and shocks lead to loss of livelihood and impoverishment of people.

Recent synthesis studies have captured this multi-causal nature of migration in Bangladesh. In the the EACH-FOR project that projected environmental change and environmental migration scenarios as well as the Foresight report, Bangladesh appeared prominently as a case study of climate- and environment-related migration. One of the EACH-FOR study papers noted an increase in rainfall variability throughout the April–October season, and a shift to a bimodal distribution pattern instead of the more common single-peak distribution, and a reduction of overall rainfall and intense rainfall in October (Ahmad et al. 2012). The rainfall variability disproportionately affects poor farmers with small land holdings and fishers by changing flood patterns (Ahmad et al. 2012).

A more recent paper based on the EACH-FOR research points out that migration has become a main coping strategy for poor households, but with high social costs (Warner and Afifi 2014). Based on participatory research, a 1300-household survey and semi-structured interviews in the drought-prone Kurigram district of Bangladesh, Warner and Afifi (2014: 5) note that migration is a major risk management and "coping strategy" in the face of environmental and climatic uncertainties such as rainfall variability and economic disadvantages. Over a third of the households surveyed had noted longer dry spells and frequent droughts as a 'very important' reason to migrate – landless, low-skilled and poor households being the most affected. Such migration could enhance food security, or be erosive, that is disadvantageous, even leading to more food insecurity (Warner and Afifi 2014). Household members often migrate to cities to cope with the impact of environmental events, even though 'environmental migration' as such is often indistinguishable, as human movements are multi-causal anyway (Foresight 2011).

When households send a member to work temporarily in towns in response to diminished farm productivity caused by riverbank erosion, soil salinity or environmental stresses and shocks such as cyclones or floods, their migration routes follow established pathways; and it is often the less poor and the more educated that migrate after such events (Foresight 2011). Studies in Bangladesh and elsewhere have shown that migration in relation to climatic reasons and environmental reasons is often connected with economic migration, and follows similar routes and makes use of established social networks (Bilsborrow and Okoth-Ogendo 1992 Warner et al. 2013: 17 Munshi 2003; Lu et al. 2012).

While climate- and environment-related migration can be advantageous or disadvantageous as described above, Foresight (2011) cautions that non-movement or some forms of movement to vulnerable places could enhance vulnerabilities.

People affected by climatic stresses and shocks are likely to get trapped in low-lying urban areas in mega-deltas and slums in growing cities with water shortages. Even while storm shelters and early warning mechanisms help save lives in the event of cyclones and floods, they could also encourage people to stay back in vulnerable areas, raising questions about the robustness of such systems to future shock. This dilemma in the context of 'disaster-proofing' vulnerable areas in Bangladesh has been addressed in a detailed World Bank study assessing adaption options in the country. World Bank (2010: 97) notes that safety measures such as embankments or polders could lead to more asset creation in their shadow that is deemed 'safe.' A combination of valuable built assets, and more extreme events could possibly lead to more human suffering and losses, as the Hurricane Katrina experience in New Orleans showed. In short, in highly vulnerable places, immobility could be a more serious risk compared with any form of migration.

At one end of such environment- and climate-related movement lies displacement after environmental shocks, but such movements tend to be short-term over short distances, mostly within two miles of residence in the case of riverbank erosion in Bangladesh, as Zaman (1989) shows. Several islands, such as Bhola and Hatia on the Meghna river estuary, have been facing high levels of riverbank erosion. Affected households often move over short distances, and often face multiple displacement, many landless families often ending up on embankments and on the riverbank itself (Shamsuddoha, and Chowdhury 2007; Biswas and Chowdhury 2012). Even in the case of longer term migration after stresses and shocks, 59 per cent of the movements occur within the district, and 39 per cent outside; 81 per cent of those who move out go to city centres, 13 per cent outside the country and 6 per cent to other rural districts (Gray and Mueller 2012).

Migration could be driven by livelihood stresses caused by climate- and weather-related events – it could be short-term displacement to escape inundation or migration to a village, town or city to earn a livelihood until the following cropping season. Most of the people affected in floods are landless labourers, as the farms they work in get inundated and remain uncultivable for a long period. Seasonal migration is also a regular feature for people affected by droughts, especially in northern Bengal where people escape the lean period between harvests called *monga* marked by poverty and food insecurity (Findlay and Geddes 2011). Such movements often become more common after environmental shocks and stresses, especially droughts and famines, mostly among the poor, although not necessarily the poorest who often cannot afford the costs of migration (Kniveton at al. 2009). Frequent cyclones are one of the main environmental drivers of migration. After Cyclone Aila in 2009, many people moved to other towns due to lack of working opportunities in the affected areas. Failures in cropping and shrimp farming due to salinisation could also alter migration patterns (WARPO 2006). After hazards people move to safety, and the landless among them move for income recovery (Penning-Rowsell et al. 2013). However, families prefer to stay put and migration appears to be the last resort.

More gradually developing stresses on livelihood could lead to different forms of circular migration. During times of *monga* or seasonal food scarcity and lack of employment during September to November, people migrate, especially from northern parts of Bangladesh. Landless people often end up in poverty and hunger, and every year boys and men from *monga*-affected areas migrate to cities and better-off villages (Siddiqui 2009). Large-scale, but often unplanned, shrimp farming in the southern coastal belt has led to salinisation of the soil and lower yields from rice fields (Rahman et al. 2013). Growing water stress and climate variability reduce agricultural productivity, helping to drive rural-urban migration. Besides, riverbank erosion displaces 50,000 to 200,000 people in Bangladesh every year (Islam and Hasan 2016). As it destroys farms and homes (Zaman 1989), sometimes communities get displaced several times (Hutton and Haque 2003) – for example a study by Abrar and Azad (2004) in northwest Bangladesh found that on average households have been displaced 4.6 times by riverbank erosion. In a projected scenario of climate change, there is a likelihood of low-lying parts of delta islands and coasts getting inundated and pushing the salinity line further north, causing further water stress and crop losses (Sarraf et al. 2011).

Demographic pressures make up another driver for migration. At 1237 people per square kilometre, Bangladesh has one of the highest population densities in the world (World Bank 2016). An analysis of the latest census figures (BBS 2012; Marshall and Rahman 2013) suggests that the population growth rate in the coastal region is lower than that in the more developed central areas of Bangladesh. While the explanation for this differential rate of growth is migration because of the lower economic growth of the region, there are indications that hazards such as cyclones and floods also play a role (Marshall and Rahman 2013).

Lastly, there could be socio-political reasons behind migration. Many village communities in Bangladesh are under the power of local mafias (Raillon 2010) that gain control over accretion land by using violence and make resettlement a socio-political issue (Zaman 1989). The government is trying to secure better land tenure rights for people displaced by climate-related stresses and shocks. As part of its rehabilitation initiatives, many landless and displaced people have been resettled in revenue land and char areas formed of riverine silt. Guchhogram Climate Victims Rehabilitation Project, for instance, aims "to settle the climate victims, landless, homeless, address-less and river eroded people on *khas* land or donated land" (Guchhogram 2010). This initiative has been discussed in chapter 8. *Khas* land means land owned by the government.

Often new settlers – in cities as well as villages – are intimidated by the local elite, and many have to go back to their places of origin as reports suggest. Unequal and unjust land distribution patterns further add to the vulnerability of people. In the *char*, areas where land tenure is temporary and ad hoc, poor people try to settle in newly formed landmasses. The government has acknowledged the prevalence of land-grabbing: As the 6th Plan document states:

> "There are land laws and policies to allot such land to the poor and the landless, but in actual allocation the interest of the poor is rarely preserved. The vested interest groups in both rural and urban areas are in de facto and de jure possession of these lands with the help of money and muscle. The ethnic

> people of the Chittagong Hill Tract (CHT) and other areas are losing their common property rights in land. In the cities, the slum dwellers pay high rent for staying in the slums and they remain under threat of eviction."
>
> (Planning Commission 2011: 68)

There is public opposition against land grabbing, forcing the Government to take further steps (Feldman and Geisler 2011). However, powerful *talukdars* (landholders) and *jotedars* (chieftains) often gain control over such land by force.

2.4 Conclusion

In summary, migratory movements in Bangladesh vary in terms of their drivers as well as dimensions of time and space. The analysis of existing patterns shows that while migration is largely economically driven, it could be influenced by climate and environmental factors, especially disasters. While the exposure of the country to multiple hazards makes migration an important coping and adaptation strategy as literature shows, socio-political and demographic factors also often play a role. Impacts of increased flooding, storm surges, riverbank erosion and drought often lead to livelihood stresses, and people make use of the income differentials between rural and urban livelihoods to offset the losses suffered or to gain more income and rebuild better. Disaster-related migration appears mostly to be short-term movements over short distances, compared with planned migration in search of better livelihoods. Long-term changes in weather and hazard patterns associated with climate change are expected to have implications on existing migration patterns, as emerging research shows. While migration can be an adaptation strategy in the face of climate change, lack of mobility due to resource scarcity, a false sense of security or other reasons could lead to people getting trapped in vulnerable places. There is also the risk of migrants moving to risky environments. This scenario of movement that is very prevalent amid exposure to multiple hazards calls for rigorous research into migration patterns in Bangladesh and their sensitivity to climate- and environment-related factors.

3 Apocalypse now?

Alarm, skepticism and evidence in the climate-migration debate

The literature on climate, environment and migration linkages has often been marked by neo-Malthusian overtones, linking climate change with resource shortages, leading to forced migration and even violence. Based on emerging evidence and theoretical advances over the past three decades, the book critiques this notion, questions projection of huge numbers and, instead, traces the multiple, complex ways in which social and ecological systems interface. It asserts that there would likely be an increase in migratory movements in the coming years with or without climate change as economic and development challenges promote human mobility at massive scales. This book attempts a close look at the multi-causal, multi-faceted nature of migration in Bangladesh amid climatic and environmental stresses and shocks and rapid economic development.

3.1 Scary sound bites

> "Climate change will stir 'unimaginable' refugee crisis, says military."[1]
> "The desperate exodus of the climate refugees."[2]
> "How climate change is behind the surge of migrants to Europe."[3]
> "Climate change and mass migration: a growing threat to global security."[4]

The headlines cited above come from a sample of top two pages of Google search results for "climate migration news." 'Crisis,' 'exodus,' 'surge' and 'threat' are the usual media terms used to describe climate-related migration. From the Sahel to Sinai, Rwanda to Bangladesh, changing climate is said to have fueled conflicts and migrant and refugee flows. The rhetoric of climate change leading to more migration of largely poor people and, in turn, resource scarcities triggering conflicts is very much alive in academic and policy circles (for instance, Reuveny 2007, 2008). The literature on climate, environment and migration linkages as such has often been marked by neo-Malthusian overtones, linking climate change with resource shortages, leading to forced migration and even conflicts and violence (for instance, Myers and Kent 1995; Homer-Dixon and Percival 1996). The media reflects these concerns and politicians see climate change as a clear and present threat to national and global security.

However, while concerns over the potentially devastating human impacts of climate change and the ensuing forced migration and its socio-political and economic implications still remain alive, many researchers are critical of this notion, and have sought instead to trace the multiple, complex ways in which social and ecological systems interface. Much of the literature reviewed in this chapter shows it is not people who run out of water in their wells or fish in their seas who migrate to Europe. That is because migration typically requires money, social networks and contacts in the destination country (Brown 2008). International migrants are usually not people who are deprived of their assets in a slow or sudden disaster, but those in search of a better life, and those who can afford such a search, as research shows in this chapter. It is not to say that migrants are rich – just that they are people who can afford to gather and spend enough resources to undertake long journeys to improve their income.

Migration being a largely economic phenomenon, the literature also asserts that there would likely be an increase in migratory movements in the coming years with or without climate change (Foresight 2011). Given the uncertainties of future migration patterns in terms of their scope in time, space, causes and consequences, climate change need not necessarily be a threat multiplier as it is projected to be in the media discourse that is in vogue.

This chapter focuses on recent research in the field of climate- and environment-related migration, tracing how the debate over this issue has led to new research, methods and evidence. Though much of the research on human impacts of climate change discussed here is rooted in human geography, the chapter also draws from social anthropology, sociology, climate science and ecology. The migration decision-making aspects discussed as part of the qualitative analysis are influenced by behavioural economics and social psychology. In this regard, research on migration and environmental change is becoming increasingly interdisciplinary, adding new dimensions to the field of migration studies.

An interdisciplinary approach helps this book take a close look at the multi-causal nature of migration in Bangladesh amid climatic and environmental stresses and shocks; and how people make decisions on staying or moving out of their place for short or long durations to nearby or faraway destinations. Setting the context for such an enquiry, this chapter first narrates how the concept of the environment-climate-migration nexus has evolved, especially over the past three decades. Then it moves on to explore theories of migration that take a behavioural approach to understand various facets of human mobility, including in the realm of climate and environment. As a next step, the chapter briefly explains how migration could contribute to adaptation and resilience, two concepts gaining currency of late in the UN, humanitarian and development interventions, especially in the context of climate change. Then the chapter reviews recent literature, including synthesis studies that present evidence for climate- and environment-related migration and its policy implications. The broad range of literature considered in this chapter sets the scene for the following chapter 3, a case study of Bangladesh that looks more closely at country-specific evidence.

3.2 Environment-climate-migration nexus: the debate

3.2.1 Early references

Historically, environmental and climatic factors have often been cited as influencing human mobility patterns. Indeed, environmental and climatic dimensions of migration gained research attention as far as a century ago. Ravenstein (1885: 286), for instance, wrote about "unattractive climate . . . producing currents of migration." However, while changes in the environment give rise to livelihood pressures and safety concerns, people's aspirations for a better life also contribute to their migration decisions.

One example for such mixed motives of migration that has gained research attention is the 1930s Dust Bowl phenomenon in the US. The southwestern Great Plains of the US, a semi-arid grassland, became home to thousands of settlers after the Homestead Act of 1862. The farmers turned millions of acres of prairie grasslands into wheat farms and pastures. However, dryland farming and overgrazing laid bare the land, and with the onset of a drought in 1930, strong winds began to blow away the top soil into dust clouds, darkening the sky, covering barns and homes, destroying farms and crops. Families in the Great Plains that stretch from Canada in the north to Mexico in the south abandoned their farms and migrated westward, in search of work (Library of Congress).

As John Steinbeck described in his famous novel, *Grapes of Wrath*:

> "And then the dispossessed were drawn west – from Kansas, Oklahoma, Texas, New Mexico; from Nevada and Arkansas, families, tribes, dusted out, tractored out. Car-loads, caravans, homeless and hungry; twenty thousand and fifty thousand and a hundred thousand and two hundred thousand. They streamed over the mountains, hungry and restless – restless as ants, scurrying to find work to do – to lift, to push, to pick, to cut – anything, any burden to bear, for food. The kids are hungry. We got no place to live."
>
> (Steinbeck 1992)

As a case illustrating the multi-causal nature of migration, the Dust Bowl offers many lessons. Narratives (for instance McWillams 1942; Johnson 1947), historical analyses (Worster 1979, 1986) and photo documentation have vividly captured different dimensions of climate-environment-human interface involved in this dramatic episode in the history of the US. Dorothea Lange's photographs taken during the 1930s for the Farm Security Administration, for instance, poignantly captured the human drama involved in the phenomenon (FSA).

The Dust Bowl was a cumulative effect of environmental and economic crises (Obokata et al. 2014) leading to migration of three million people (Boano et al. 2007). This migration included rural-to-urban, urban-to-rural and rural-rural movements (McLeman et al. 2014). Researchers have approached this event from different angles. One viewpoint is that settlers who came in the late 19th and early

20th centuries converted large swaths of grasslands into grain, corn and cotton farms. The prevailing economic ethos promoted over-exploitation of nature for profit, with farmers underestimating the risk of drought (Johnson 1947; Worster 1979). According to this perspective, converting grassland to wheat, combined with the great drought of the 1930s, led to massive dust storms and mass migration. This point of view has influenced later analyses of socio-economic processes and hazard risk profiles (for instance, Blaikie et al. 1994; Oliver-Smith 1996). Blaikie et al. (1994) argue that social, political and economic contexts cause disasters as much as the natural environment. Oliver-Smith (2013) points out that the Dust Bowl migration to California was an outcome of the Great Depression of the 1930s as much as the drought. He draws parallels between the 1927 great flood of Mississippi that displaced 700,000 people (including 330,000 African Americans), Dust Bowl migration to California and Hurricane Katrina of 2005. In these cases, flood, drought and hurricane, respectively, worsened the prevailing socio-economic conditions and labour practices, triggering mass migrations.

However, later studies using newer tools – including economic datasets and maps – have provided new insights into the Dust Bowl phenomenon. Some of the studies questioned the notion that it was an ecological failure led by profit maximisation. Cunfer (2005), for instance, countered the notion of over-farming, showing that the balance between cropland-pasture remained virtually stable form the 1920s until the 1990s. More recent studies, especially in the wake of Hurricane Katrina in 2005 and the financial crisis in 2008 leading to migration have given newer insights into the relationship between environment and population movements. One such finding is that it was not necessarily settled farmers who migrated after all. The farmers were learning to adjust and adapt to local conditions, and the more experienced among them took care of the land; while later arrivals often engaged in badly maintained monoculture grain farms, from which dry soil drifted across, making more farms barren (McLeman et al. 2013).

Another interpretation is that more public spending as work relief, public works, direct relief and social aid reduced migration from the areas that were benefitted; but attracted migration from elsewhere (Fishback et al. 2006). Therefore, the argument is that migration during the 1930s was lesser than during the decades before and after; and had the New Deal spending (to offset the impact of the Great Depression) been more evenly distributed across counties, it would have had still lesser impact on net migration (Fishback et al. 2006). These interpretations of the Dust Bowl phenomenon show the multi-causal nature of environment-related migration and challenges it poses in terms of understanding and responding.

Despite such a widely talked about environmental event such as the Dust Bowl, the human-environment interface often remained a mere backdrop rather than a core concern in migration research (Hunter 2005; Piguet 2010) – until a certain point when the situation changed dramatically. As Oliver-Smith (2012) argues, underplaying of the environmental factors in studies related to migration could have been due to the dualistic western notions that tend to place nature as something distinct from human beings and their societies. That meant there was little solid evidence to show to what extent environmental factors influenced migration. It may be noted that even though environment was not the key focus of empirical studies, theorists still included environmental factors as drivers of movement, as explained in section 3.3 of this chapter. The point is that there was little effort to isolate environmental and climatic factors from a set of other drivers of migration, such as population increase, poverty, conflicts and better income prospects.

3.2.2 Drummed-up fears of climate migration

Even with a thin empirical base, the concept of the environment-migration nexus rather dramatically began to gain currency during the 1970s and 80s. This happened in the background of the growth of the environmental movement, increasing scientific evidence for global warming (Peterson et al. 2008) and climate change becoming a global policy concern. An early example of this period noted: "As human and livestock populations retreat before the expanding desert, these 'ecological' refugees create even greater pressure on new fringe areas, exacerbate the process of land degradation" (Brown et al. 1976: 39). Later research projected a grim scenario of forced migrants in their millions putting undue pressure on the environment, in some cases violently competing over resources, in an apocalyptic backdrop of climate change (El-Hinnawi 1985; Jacobson 1988; Tickell 1989; Myers 1993; Myers and Kent 1995; Homer-Dixon and Percival 1996; Myers 2001).

Against a background of population growth in poor countries, environmental events and processes could lead to shortage of resources, conflicts and displacement, as the researchers mentioned above projected. Internal as well as international movement involved a projected 50 million people by 2010 (UNFCCC 2007) to 250 million by 2050 (Christian Aid 2009). Follow up reports were published even in the noughties (for instance, Stern 2007; Biermann and Boas 2010). Often newer reports built on or repeated earlier figures, especially the ones used by Myers and Kent (1995) (for instance, Christian Aid 2009; Oxfam 2009). In a review, Jakobeit and Methmann (2012) listed a set of such recent reports, many of them recycling old figures as table 2.1 shows. As Gemenne (2011: S41) puts it in an article title, "the numbers don't add up," with critics dubbing them "artificially inflated, excessively alarmist, or 'guesstimates.'"

It may be noted that the discourse was not about the rather uncertain and nuanced phenomenon of migration related to climate or environment, but about the rather unproven certainties of 'environmental refugees' (El-Hinnawi 1985), 'climate refugees' (Bierman and Boas 2010) or 'environmental exodus' (Myers and Kent 1995). The first UN intergovernmental report on climate change, for instance, noted: "The gravest effects of climate change may be those on human migration as millions will be displaced" (IPCC 1990: 20). UNHCR (1993) acknowledged what it called "clear links" between degradation of the environment and refugee movements, asserting that the deterioration of the natural resources in a background of demographic pressure and chronic poverty could trigger or amplify political, ethnic, social and economic tensions. Even though both of these organisations have since taken much more nuanced positions on the subject, many other international bodies, UN organisations and NGOs have repeated these concerns.

Table 3.1 Various projections of climate migrants and refugees

Report	*Projected year*	*Projected figure*
Christian Aid (2009)*	2050	250 million
Oxfam (2009)*	2050	150 million, 75 million in Asia Pacific
Greenpeace (2008)*	2100	125 million in India-Bangladesh
Tearfund (2006)*	2050	200 million
Myers (2002)*	2050	250 million
Stern (2007)*	2050	150–200 million
UNFCCC (2007)	2010	50 million
UNEP (2008)	2010	60 million
GHF (2009)*	2009 2050	26 million + 25 million (indirect) 150–200 million
UNU-EHS (2005)	2010	50 million

(Myers and Kent 1995; Gemenne 2011; Jakobeit and Methmann 2012)

* *Refers to Myers and Kent (1995).*

3.2.3 Challenging the rhetoric of climate migration

Even while acknowledging that changes in climate and environment could drastically alter human habitats and thereby influence mobility patterns, critics have questioned the projected apocalyptic scenario of climate migration. McGregor (1994: 121), for instance, challenged the notion of "simple and direct cause-and effect link between climate change and migration" and argued that the idea of environmental push factors was inadequate to explain migration. Instead, people's entitlements (Sen 1981), access to resources, the role of social institutions, labour relations, culture, social networks and human agency together determine their differential vulnerability to environmental adversities and migration decisions (McGregor 1993).

In a comprehensive critique, Black (1998) challenged the rhetoric of environmental migration, citing case studies from Africa, Asia and Latin America. He looked at what appeared to be a straightforward relationship between environmental degradation and forced migration and found a lack of convincing evidence to link the two phenomena. A closer look at cases of environmental degradation often cited as a cause of migratory or refugee movements, especially in Africa, revealed that they were not necessarily related to people's movements. This finding prompted the framing of migration as an essential part of socio-economic processes rather than an outcome of environmental decline. Based on these findings, Black (1998: 182) suggested that researchers look for "spatial and temporal relationships between periods of forced migration and environmental decline" and gather better evidence for climate change-related phenomena such as sea level rise. Echoing similar concerns in a review article, Lonergan (1998) noted that though environmental degradation and resource depletion may contribute to population movements, the linkages were embedded in a matrix of poverty and inequity. Migration attributed to environmental causes was found to be a form of coping mechanism that people have been practicing historically (Black 2001).

Later Castles (2002) commented that environmental factors are always closely linked to political and economic factors, including globalisation and exclusion of whole regions from global development, leading to decline in living standards.

Clearly, a divide between those forecasting waves of 'environmental refugees' and those adopting a more skeptical stance became evident by the 1990s. Recognising such a sharp divide, Dun and Gemenne (2008: 10) call researchers who isolate environmental and climatic factors as the main driving forces of migration "alarmists," and those who stress the complexity of the migration process "sceptics." The sceptics challenged the "maximalist" (Suhrke 1994: 4) figures and the simplistic models derived from "common sense" (Perch-Nielsen et al. 2008: 375) behind these projections of mass migration. Critics also challenged the empirical evidence, alarmist tone and the stress on environmental change as the sole or main reason for large-scale human movements (McGregor 1993; Black 2001; Castles 2002). They spotted considerable data problems in studies linking demographic processes with the rural environment in development countries, and demanded more rigorous research (Bilsborrow 1992; Suhrke 1994; Hugo 1996). One suggestion put forward by several of these critics is to take into account social, economic and political factors while looking at environment-related migration.

A new academic consensus was beginning to emerge by the 1990s, acknowledging that environmental factors cannot be studied in isolation to understand people's movements. At the same time, different concepts in the environment-migration discourse became targets of criticism. In this regard, one contentious issue that still sparks debates is the term 'environmental refugees,' defined as people forced to leave their habitat because of a marked environmental disruption, including human-made changes (El-Hinnawi 1985). Scholars rejected the term on political, legal and ethical grounds (Kibreab 1997; Black 2001; Castles 2002; Renaud et al. 2011). Reviewing literature on 'environmental refugees' Black (2001) argued that concern about poor people leaving fragile environments has not translated into hard evidence for what causes their movements, or any real theoretical or empirical insight. Several streams in the continuing research in the field of climate, environment and migration take forward this quest for evidence, as section 2.6 of this chapter shows.

As the literature discussed above shows, a combination of harsh or deteriorating environmental and living conditions have historically driven migration (Black 2001). While sudden shocks might involve distress migration or displacement at a local level, changes over long periods also change people's lifestyles and livelihoods in specific geographies (McLeman 2014). Response to rising vulnerability involves a layered and complex social system across different scales of time and space, and permanent relocation is typically a least-favoured, last-resort option (McLeman 2014). In this context, the very logic of attributing a mono-causal relationship between climate change and migration in research as well as policymaking could be problematic (Nicholson 2014). Instead, as recent studies show, human movements can be seen as multi-causal processes with economic, social, political and cultural dimensions (Perch-Nielsen et al. 2008; Barnett and Webber 2010; Black et al. 2011a)

3.3 Migration as a behavioural response to changes in the environment

3.3.1 Building blocks of theory

Even as migration involves multiple drivers, people's decisions to stay or leave are influenced by a huge range of factors (Black 2001). Migration theories have sought to understand why people leave or stay by studying the behavioural and other factors involved in migration. This section first traces how different strands of theory view migration, and what role they ascribe to climate. While a behavioural approach in geography tries to understand people's interactions with environment by understanding human behaviour, cognitive research looks more closely at decision-making at an individual level – still explaining its impacts on and influence from multiple hierarchies, i.e., households, communities, districts and so on.

These enquiries build on behavioural approaches, but more closely look at socio-cognitive variables that influence perceptions, beliefs, motivation and their decision-making patterns under uncertainties (Grothmann and Patt 2005). For instance, in Bangladesh, many people find themselves helpless in the face of floods and cyclones they believe to be acts of God (Schmuck 2000). Such beliefs influence the way they respond to disasters. This book goes beyond the behavioural approach in migration research and moves into the realm of cognitive analysis to understand people's responses to climatic stimuli (Grothmann and Patt 2005; Kuruppu 2009; Kniveton et al. 2011, 2012; Reckien et al. 2013). This line of research draws from behavioural economics and social psychology (e.g., Ajzen and Fishbein 1980; Tversky and Kahneman 1991), as Grothmann and Patt (2003) show.

3.3.2 Migration as a part of the modernisation process

The question that has puzzled social scientists over the years is a rather simple one: why do people migrate? There have been many explanations for what makes people move around across different spans of time and space. Here come migration theories. Migration theories can be classified into two broad paradigms – functionalist and historical-structuralist (de Haas 2010). Functionalist theories consider migration as something that benefits the society in general by reducing inequalities within it and in relation to other societies (Castles et al. 2014). The notion that unequal income and opportunities drive migration from low- to high-income areas has been a dominant theme in migration studies (for instance, Ravenstein 1885). Simply put, these models assume that societies try to attain an equilibrium, as various economic, environmental and demographic factors push migrants out of their places of origin and pull them into new destinations (de Haas 2010). Following up on these push and pull theories, neoclassical migration theories (for instance, Todaro 1969; Harris and Todaro 1970) give a more sophisticated explanation for people's migration from villages to cities. They frame migration as an individual choice, driven by a desire to maximise income (Massey 1993).

These theories are based on neoclassical economics that considers wage differentials and other employment benefits across different work settings. Individuals leverage from these geographical differences in the supply and demand for labour by migrating to places that have a shortage of labour force and/or better income prospects (de Haas 2007). In this framing, people migrate from places with less income to places with a shortage of labour force. To be more precise, their movements are explained by spatial differences in the relative scarcity of labour and capital. So it follows that migration is not something that just happens when climate changes and resources dwindle, as it is often made out to be, especially, but not exclusively, in the tabloid media. Migrations involve a decision-making process by agents, which means rational actors, as social scientists explain (not travel agents, by the way). Thus, migration becomes an essential part of the modernisation process. It is migrant labourers who build and run big, growing cities, not always sons of the soil. Growing urban-centred economic activities encourage more and more people to move from villages to cities in search of better income (Todaro 1976; Skeldon 1997).

According to neoclassical theories, migration optimises the allocation of production factors such as capital and labour, leading to better efficiency and equilibrium. While this book acknowledges the economic aspect of migration as spelt out in the neoclassical approach, it still tests the limits of the framework. It argues that even in the context of urban growth and better income prospects in cities, and in the face of biting poverty and extreme weather and environmental events that reduce productivity and security back home, people often do not or cannot migrate. Even when there is a clear need and opportunity to gain better income by migrating, often people are unable or unwilling to migrate, as this book argues, and such limits and decisions are often linked to environmental and climatic factors. It explores the extent to which interrelated environmental and economic factors drive and sometimes prevent migration, and considers the finer cognitive elements or, simply put, the black box of migration decision-making.

On another plane, however, the notion that migration involves free choice itself has been debated. This debate is built on structuralism. Structuralism involves a complex set of theories, but it can perhaps be summarized in British philosopher Simon Blackburn's words as "the belief that phenomena of human life are not intelligible except through their interrelations. These relations constitute a structure, and behind local variations in the surface phenomena there are constant laws of abstract culture" (Blackburn 2008). Historical-structuralists, for instance, have argued that free individual choice does not exist, as structural forces limit people's options; therefore, people have to migrate when local political economies are changed by global forces (de Haas 2007).

Rooted in neo-Marxist economic theory, historical-structuralists take into account socio-economic, political and cultural settings and structures that constrain and influence individual behaviour. People are often forced to migrate when their traditional economic structures are undermined, as they become part of the global political-economic system (Castles et al. 2014). Global inequalities in income, political freedom and quality of life influence migration, and it becomes an intrinsic part of development, globalisation and social transformation. (Castles et al. 2014). In this framing, migration is more of a patterned phenomenon, with market forces, social stratification and structural inequalities restricting people's choices. Migration, therefore, is often explained as an outcome of disruptions caused in the process of capitalist accumulation, a manifestation of capitalist expansion with its inherent inequalities (Massey et al. 1998). As opposed to neoclassical approaches, from a structuralist viewpoint, migration is seen as a process that further deepens inequalities across regions and within the society.

This book does not take an explicitly historical-structuralist approach. However, it questions the notion of free choice involved in migration. It argues that often poor people are often unable to move out of hazard-prone places because of lack of resources or exposure; some hazards such as riverbank erosion and frequent floods further impoverish them and limit their ability to migrate. In the context of climate- and environment-related migration, immobility is less studied. Still it stretches the limits of migration theories. It is counterintuitive given that migration is seen as an activity that people undertake to improve their income. Here the argument is that sometimes the income levels are driven so low by environmental conditions such as hazard exposure that people cannot afford to migrate.

While it is impractical to adopt a 'grand theory' that encompasses all aspects and types of migration (Castles et al. 2014), various facets of migration have been explained by newer theories of migration. The book draws from some of these theoretical contributions. Many assumptions of neoclassical theory, such as individual agency, have been critiqued in the New Economics of Labour Migration (NELM) theory (Stark and Bloom 1985; Stark 1991). NELM acknowledges that migration decisions are not made by individuals acting alone, but by groups of related people, especially families and households, in an effort not only to maximise income, but also to minimise risks and offset the impacts of losses.

In an NELM framing, migrants do not act alone, but institutions such as families influence them to maximise expected income, minimise risks, offset market losses and leverage labour opportunities (Stark and Levhari 1982; Stark 1984; Stark and Bloom 1985; Massey et al. 1993). It could be a household-level decision to improve income by sending one or more members to a different place or shift residence altogether, an effort to gain social mobility or a search for a better and safer place to live. In these cases, the decision-making process is determined by expectations of rewards in a new location (De Jong and Fawcett 1981). This involves diversification of resources, including labour by different family members. One or more family members may migrate and send remittances to support the others who stay back at home so that even if local economic conditions and livelihoods fail or deteriorate, remittances can act as an insurance and supplement (Massey 1993). Such a framing of sharing and offsetting risks at a household level is especially useful in the case of climate- and environment-related migration. This book acknowledges the role of households in migration decision-making.

3.3.2 The role of the environment and climate in migration decisions

Migration theories based on neoclassical economics have, often, addressed the role that environment plays in migration decision-making, along with other factors such as climate, perception of changes in the environment, as well as socio-economic factors, including networks (for instance, Wolpert 1965, 1966). Wolpert's (1966: 93) stress-tolerance model of migration, for instance, defined stress as a set of "noxious" environmental forces and strain as an individual's reaction to it. Stress could influence decisions related to staying or moving, though non-movement is not considered an "equilibrium position." On similar lines, Brown and Moore (1970), for instance, theorised environment as a source of stimuli, some of the stressors that can disrupt or threaten household behaviour patterns.

However, this influence of the environment can have differential impact on individuals based on how each one of them perceives the stress, what they know about alternatives, and the way they respond (Brown and Moore 1970). Wolpert (1965: 161) argued that population movement is an interaction over space but the points of origin and destination attain significance only in the way they are perceived by the active agents. He made the distinction between the objective stimuli and the perceived stimuli to which an individual reacts while making the "mover-stayer decision" based on something called "bounded rationality." Bounded rationality is a concept advanced by Simon (1996) to explain the decision-making process within the constraints of information, ability to grasp the information and time. It suggests that the rationality of decision-making is limited by the available information that people have, the cognitive limitations of their minds and the time available to make the decision. The search and selection of alternatives and the decision to relocate or to adjust to one's current location depends on the information that he or she has about future locations. This information is often gathered from the media, friends, relatives and employment agents (Brown and David 1970), as we will see from stories of people in Bangladesh villages.

In this context, Wolpert (1965: 161) viewed migration "as a form of individual or group adaptation to perceived changes in environment, a recognition of marginality with respect to a stationary position, and a flow reflecting an appraisal by a potential migrant of his present site as opposed to a number of other potential sites." However, other forms of adaptation, such as changed farming practices, for instance, were seen to be more common than change of residence and livelihood activity. The notion of movement as a way of adaptation has become a highlight in recent literature in the field of climate- and environment-related migration (for instance, Barnett and Webber 2010). In the context of climate change, adaptation has been defined as an "[a]djustment in natural or human systems in response to actual or expected climatic stimuli or their effects, which moderates harm or exploits beneficial opportunities" (Parry et al. 2007: 6). While higher temperatures lead to sea level rise, and extreme weather events and uncertainties subsequently affect natural resources, assets and, in turn, livelihoods, safety and well-being of people (McCarthy et al. 2001), adaptation becomes a necessity. The role of migration as a climate change adaptation measure has been discussed in detail in section 2.4.

Among models of migration behaviour at macro and micro levels, an influential contribution is the value-expectancy model (De Jong and Fawcett 1981) that explains individual, household and societal determinants of migration. In this model, motivation is defined as a function of the value placed on certain goals and the perceived likelihood that a behaviour will lead to those goals. Such goals include comfort, along with wealth, status and autonomy. Comfort means a pleasant, healthier or less stressful living environment. As follow up work has shown, people respond to environmental pressures by changing behavioural factors as well as social relations (Bilsborrow and Okoth-Ogendo 1992), mediating land-use practices and such responses. While the utilitarian value of place and wage differentials across geographies contributes to the push and pull for migration, the migrants also take into account environmental factors, expectations of modernisation as well as social considerations in their decision-making process.

Migration may not always be aimed at securing a better job or even comparatively better living conditions, it could be just an escape from temporary disruptions or a strategy to offset a reduction in income back home (Stark and Levhari 1982). Extreme forms of such disruptions, namely disasters, could lead to forced migration. Oliver-Smith (1996: 303) defines a disaster "as a process/event involving the combination of a potentially destructive agent(s) from the natural and/or technological environment and a population in a socially and technologically produced condition of vulnerability." In this framing, disasters are symptomatic of the failure of a society to adapt well with its natural as well as socially constructed environment (Oliver-Smith 1996). Hurricane Mitch of 1998, for instance, overwhelmed and disrupted social, economic and environmental processes of Florida (Oliver-Smith 2009). Forced migration from such disasters and environmental crises results from interactions among climatic, environmental and social systems that lead to the catastrophic event, blown up even bigger by the event itself – so focusing on the environmental factors as the main or sole cause of migration is a fallacy, overshadowing socio-political and economic aspects (Black 1998, 2001, 2011a; Oliver-Smith 2012).

As opposed to viewing climate-related migration as a desperate act of helpless people (as some of the literature cited in section 2.2 shows), the above studies show that at varying degrees, migration involves a decision-making process to make the future better or to escape from a condition that involves risk or losses. A number of factors mediate this process, as the literature further shows. These factors include variability in the climate and the environment as well as characteristics of the migrants, at what stage they are in their life cycle, demographic features, as well as socio-economic status (Wolpert 1965). It may be argued that the framing of environmental change as a phenomenon that influences existing multiple drivers of migration (Black et al. 2011a; Foresight 2011) is a logical follow up of neoclassical migration frameworks made more relevant in the context of current concern about climate change.

Migration decisions are often unique to individuals, as further research in this field notes. These decisions depend on a series of intervening factors and personal and household characteristics. These factors work at an individual level, through local networks or at a macro level – which could include the global effects of climate change (Schmidt-Verkerk 2011). Some of the theories of migration acknowledge that individual values and attitudes (Ritchey 1976), feelings and an exercise of independent will or agency (Stark and Bloom 1985) play a role in migration decision-making. The social structure of communities, and individual migrants' characteristics and status within the structure also have an effect on migration (Ritchey 1976). A combined reading of the literature cited above shows that environment- and climate-related migration involves multiple causes, influences and individual agency, mediated by social norms, traditions and cultural backgrounds. As Kniveton et al. (2011: S35) put it succinctly, each migration decision is driven by a "unique combination of experiences, biases, assets and perceptions."

3.3 Taking a deeper look into mental processes

Just as migration decisions are unique, the notions of risks that make people take these decisions are also unique. Social and cultural factors determine the way people perceive and accept risks (Slovic 1987). People often misjudge risks and natural hazards. Flood plain dwellers, for instance, have difficulty in assessing the probability of a major flood. Many people tend to think that a major flood might not occur soon after another one. Often they replace the uncertainty of such hazard events by seeing them as cyclical, repetitive occurrences or apply a law of averages that rules out major events one after another (Slovic 2000). Besides, beliefs and perceptions of risks are related to the way people experience natural hazards and the culture and livelihoods that communities share. Besides, people often choose or feel compelled to live in dangerous places because of livelihood pressures (Cannon et al. 2014). Therefore, to learn how different communities and individuals respond differently to climate- and environment-related stimuli, a close look into mental processes that shape perceptions, experiences and responses of climatic stresses and shocks is needed (Grothmann and Patt 2005). This section deals with literature that offers such a close look.

3.3.1 Towards cognitive analysis

While a behavioural approach in geography tries to understand people's interactions with environment by understanding human behaviour, cognitive research looks more closely at decision-making at an individual level – still explaining its impacts on and influence from multiple hierarchies, i.e., households, communities, districts and so on. Such enquiries build on behavioural approaches, but more closely look at socio-cognitive variables that influence perceptions, beliefs, motivation and their decision-making patterns under uncertainties (Grothmann and Patt 2005). For instance, in Bangladesh, many people find themselves helpless in the face of floods and cyclones they believe to be acts of God (Schmuck 2000). Such beliefs influence the way they respond to disasters. This book goes beyond the behavioural approach in migration research and uses moving into the realm of cognitive analysis to understand people's responses to climatic stimuli (Grothmann and Patt 2005, Kuruppu 2009; Kniveton et al. 2011, 2012; Reckien et al. 2013). This line of research draws from behavioural economics and social psychology (e.g., Ajzen and Fishbein 1980; Tversky and Kahneman 1991), as Grothmann and Patt (2003) show.

In this field, the literature has shown that climate risk perception and perceived adaptive capacity – what an individual or a community thinks it can do, given the availability and access to resources – are as important as objective adaptive capacity, or what they can actually do (Grothmann and Patt 2005). They are among the determinants of the adaptation decision-making process. Adaptive capacity means the "ability of a system to adjust to climate change (including climate variability and extremes) to moderate potential damages, to take advantage of opportunities, or to cope with the consequences" (IPCC 2007a).

In migration studies, De Jong et al. (1986) support the Ajzen and Fishbein (1980) theory of reasoned action to a large extent. They have included among explanatory variables, migration intentions, behaviour, family pressure – to move or stay – family networks at alternative destinations, resources to move, prior migration experience and the life cycle stage, including marital status and age. Personal and structural backgrounds were shown to have an independent and direct effect on migration behaviour. By extending this theory to incorporate the additional parameter of perceived behavioural control, Ajzen (1991) created the theory of planned behaviour. Intended to aid prediction of behaviours over which a person does not have complete voluntary control, perceived behavioural control is conceptualised as the expected ease of actually performing the intended behaviour.

Including attitudes towards a behaviour, a subjective norm and perceived behavioural control (as well as the beliefs held by an individual that make up these components), the theory of planned behaviour can be used to effectively break down the reasoning process relating to the development of a behavioural intention to make migration decisions. Thus, within the theory of planned behaviour, the intention to perform a behaviour is considered a direct antecedent of the behaviour. Attitudes are thought to represent an evaluation of the perceived consequences of behaviour and likelihood of outcomes, whereas social norms can be considered as accepted standards conveyed by peers, family, community or society.

A major facilitating factor for attaining perceived behavioural control is having previously engaged in the behaviour (De Jong 2000). De Jong (2000: 318) further argues that "intentions, expectations, norms, and gender roles are key elements inside the black box of migration decision making." The approach for this book is rooted in the theory of reasoned action and the theory of planned behaviour.

In an influential work that probes the decision-making process, Grothmann and Patt (2005) offer two case studies from urban Germany and rural Zimbabwe to explain the cognitive influence of adaptive action. Their Model of Private Proactive Adaptation to Climate Change (MPPACC) separates out the psychological steps involved in taking action in response to perception of changes in the environment, and examines factors that hinder or promote adaptive action (figure 3.2).

The model takes into account risk perception and perceived adaptive capacity, largely neglected in earlier research. Smith et al. (2010) further developed the model to explore the nexus of migration and climate change. The Model of Migration Adaptation to Rainfall Change (MARC) proposed by Smith et al. (2010) and further developed in Kniveton et al. (2011) builds on MPPACC and seeks to explain how individuals decide to migrate when the rainfall patterns change (figure 3.1).

The MARC model identifies risk perception and perceived adaptive capacity as key factors in this regard and represents individual migration decision-making and related input components that shape an agent's decision to migrate under changing rainfall conditions (see figure 3.1). This model is divided into four hierarchies: structural, institutional, individual and household. This conceptual approach takes its theoretical basis from the social psychological theories of reasoned action and planned behaviour (Grothmann and Patt 2003) and the follow up work in Grothmann and Patt (2005).

Practically, MARC is an agent-based model (ABM), used to simulate future large-scale migration behaviour. The model incorporates related components such as social and environmental inputs that shape individuals' decision to migrate at four levels – structural, institutional, individual and household. Such studies promise process-based models of behaviour that are valid across cultures, and recognise the multi-causality of migration decision-making and complexities and uncertainties involved in developing decisions.

The cognitive enquiries, however, tend to assume that people analytically assess and calculate the desirability and likelihood of possible outcomes, thereby underplaying or ignoring feelings about objects, ideas, choices, mental images and emotions. Theories of choice under risk or uncertainty tend to be cognitive and consequentialist, using rational choice models (Leiserowitz 2006). However,

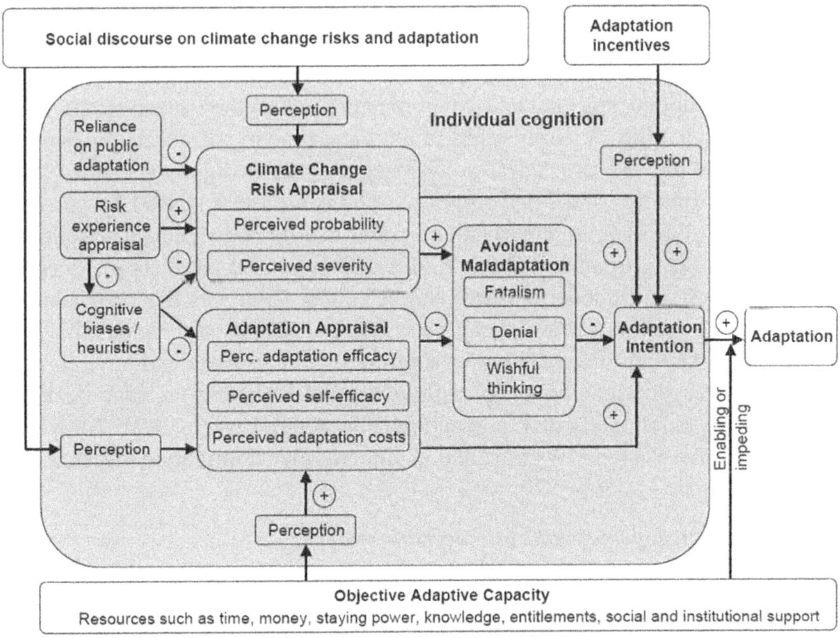

Figure 3.1 Model of Private Proactive Adaptation to Climate Change (MPPACC)

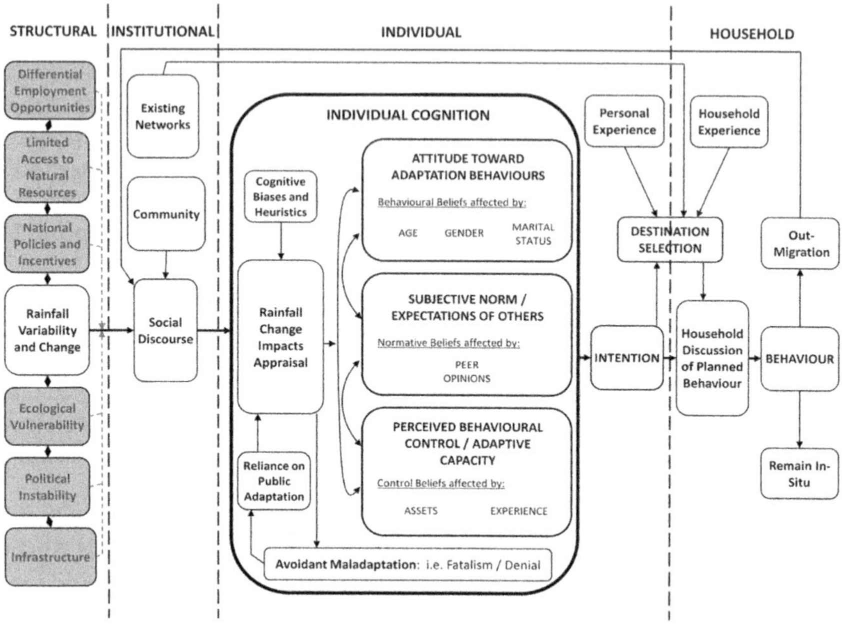

Figure 3.2 Conceptual model of Migration Adaptation to Rainfall Change (MARC)

a closer look reveals the subjective nature of climate experience, as climate is closely connected to the identity of people in a particular place.

People's decisions to deal with risk are not always rational, but linked to the emotional, symbolic, spiritual and environmental values they attach to the place and people's identities. Identity refers at the same time to social categories of an individual and to the sources of his or her self-respect or dignity (Fearon 1999). Perceptions of climate risks and response options are often shaped by personal observations of a changing environment, belief in God and significance of home and responses (Mortreux and Barnett 2009). The cultural products of a certain place often reflect the local climate. Hulme (2008: 7) argues that "registers of climate can be read in memory, behaviour, text and identity as much as they can be measured through meteorology." So in the face of climate and environmental change, decision-making is not a fully rational, deliberative, analytical act, but rather an "emotionally driven experiential system" (Epstein 1994: 709). This route of enquiry is a road less travelled in climate-related migration research.

3.3.2 Experience, perceptions and decision-making

Though the subjective nature of the decision-making to adapt (or migrate) is rarely studied, on a related track, Kuruppu and Liverman (2011) built on Grothmann and Patt's (2005) model, adding affective heuristics, intention implementation plan development and the stages of change. Heuristics are mental toolkits for decision-making and problem-solving; and an affective heuristic involves emotions influencing the cognitive processes. In their study in *Kiribati*, Kuruppu and Liverman (2011) have noted that people wanted to adapt better with more effective water management practices when they perceived climate change as a process that they could feel and relate to; and the more the people believed in their own capabilities, the more they wanted to take up such measures. Though cultural practices – including traditional knowledge and rituals – may influence objective adaptive capacity while people pursue coping or adaptation strategies (Kuruppu 2009), governments and development agencies often do not take them into account when planning adaptation strategies (Kuruppu and Willie l. 2014).

On the flipside, emerging research shows that it is not only resource constraints and socio-economic factors that limit adaptation choices, but also psychological factors, habits and perceptions of climate variability (Dang et al. 2014). Therefore, migration scholars note that barriers to movement could be internal as well as external – people may not be able to move or in some situations they just do not want to move despite the risks involved in staying at a vulnerable place. Black and Collyer (2014: 52) argue that distinguishing between these two scenarios could be extremely difficult, requiring a "nuanced reframing of migration theory" concerning "migratory space, local assets and cumulative causation."

In terms of cumulative causation, climatic stresses and shocks influence drivers of migration (Black et al. 2011a; Foresight 2011); and these drivers have individual, household, environmental and structural dimensions (Kniveton et al. 2011). Furthermore, making decisions in multi-faceted and complex environments benefits from non-conscious processes in response to experiential learning during repeated exposure to novel situations, ideas and relationships (Beratan 2007). It follows that environmental changes and climate variability are entwined with livelihood opportunities and limitations, people's experience and perceptions of climate and environment and their expectations from alternative courses of action.

Empirical studies have tested how perceptions and expectations shape migration decisions along with socio-economic factors. De Jong's (2000: 307) logistic regression models using longitudinal data from the 1992 and 1994 waves of the Thailand National Migration Survey shows that "a strikingly different set of expectations, household demographic indicators, and migrant capital factors were significant determinants of migration intentions for men and women." Intentions could be a predictor of more permanent migration behaviour, and expectancies about achieving future goals are determinants of intentions to move for both men and women. However, gender roles, including marital status and responsibility to take care of dependents, are important determinants of migration intentions for both men and women. Prior migration experience as an indicator of direct behavioural control is a strong predictor of both migration intentions and behaviour (De Jong 2000). Furthermore, De Jong's (2000) research shows that determinants of migration intentions and migration behaviour are not the same.

Taking into account socio-cognitive variables of climate risk perception and perceived adaptive capacity as key determinants of the adaptation decision-making process, Grothmann and Patt (2003) see adaptation as a socio-cognitive-behavioral process. They explain "adaptation not only as adaptive behavior, but also as changes in cognition (e.g., risk perceptions), which are socially constructed and negotiated. For example, the behavioral adaptation of communities in flood-prone areas to an increased risk of flooding due to climate change (e.g. by building higher levies or houses less prone to damage by water) is preceded by an increase in perceptions of the risk of flooding" (Grothmann and Patt 2003: 3). This book adds the element of migration as an adaptation strategy.

3.4 Gaining resilience and adapting to climate change

3.4.1 Bouncing back, modifying, adjusting . . .

Though the focus of this book is on determinants of migration, it draws from research on resilience and adaptation to see whether migration in the context of climatic and environmental hazards qualifies as adaptation. The rationale of this approach is that the respondents who have migrated have reported (as chapter 4 shows) that they found it a beneficial activity that helped them offset losses suffered because of hazard exposure, and better prepare for their future in risky environments. In that case, migration becomes an adaptive activity, though it is hard to measure the extent of adaption that a migrant can achieve. As such, the conceptual intertwining of migration with other societal and contextual processes comes from a socio-ecological systems approach. In this framing, social patterns of behaviour are seen to co-evolve in an environment that changes its physical, demographic, economic, social and political characters (Rammel et al. 2007; Kniveton et al. 2012). The literature shows migration often becomes foremost a livelihood strategy undertaken amid multiple opportunities, stresses, shocks and, above all, uncertainties – an activity interwoven with other societal processes (McLeman and Smit 2006). Therefore, migration can be seen as a good strategy for adaptation to environmental change, an "extremely effective" way towards gaining long-term resilience (Foresight 2011: 10).

In this regard, resilience can be defined as "a measure of the persistence of systems and of their ability to absorb change and disturbance and still maintain the same relationships between populations" (Holling 1973: 14). A later and widely cited definition of resilience calls it "the capacity to use change to better cope with the unknown; it is learning to bounce back" (Douglas and Wildawsky 1982: 196). Resilient social-ecological systems use a set of diverse mechanisms – learning from changes and shocks, thus sustaining themselves by adapting to disturbances (Adger et al. 2005). Resilience, in short, is not the opposite of vulnerability, but a more comprehensive concept, a route from disaster risk reduction to sustainability. It explains the capacity of a community to withstand the effect of a stressor; it is fundamentally a function of a sociocultural system. (Oliver-Smith 2012). It is the ability to plan and prepare for, absorb, recover from or more successfully adapt to actual or potential adverse events (National Research Council 2012).

In the context of climate change, migration exists on a continuum to maladaptation by displacement to an effective adaptive response in many ways. Maladaptation denotes "an adaptation that does not succeed in reducing vulnerability but increases it instead" (McCarthy et al. 2001: 990). As a successful adaptation strategy, migration can be a solution to challenges thrown up by climate change in habitability of a place, productivity of the farm, availability of food, water and energy and exposure to hazards, prices, job opportunities – at points of origin and destination of migration (Foresight 2011). In the context of disasters such as the 2011 drought in the Sahel and the Horn of Africa, and the 2010–2012 floods in Pakistan that have caused large-scale population movements, it has been claimed that environmental change – as well as political instability – further complicated migration patterns by adding uncertainties, vulnerability and livelihoods stresses (IOM 2013). In such contexts, mobility can help people reduce and recover from the impact of hazards, and boost their ability to access and use material assets and social networks. It can be a mutually beneficial option for the migrants as well as their hosts (Foresight 2011).

Migration can be considered a positive step if it is a livelihood strategy that can enhance people's income and generate financial and social remittances that contribute to better resilience back home (McLeman and Smit 2006; Barnett and Webber 2010; Tacoli 2009). However, promoting migration – or resilience or adaptation for that matter – is not without its share of political problems. Recent literature has critiqued migration or planned relocation or promotion of migration as a policy option for resilience and adaptation on economic as well as socio-political grounds. Shumway et al. (2014), for instance, argue that migration can help communities achieve better spatial equilibrium, or it can just amplify already unequal income distribution patterns across regions. Analysing migration and income change in the US during 2000–2010, Shumway et al. (2014) note that countries that are the worst hit by environmental hazards are losing on account of net outmigration, and because in-migrants have lesser income than those who move out. Discussing small island developing states (SIDS), Kelman (2015) argues that migration from such islands, relocating and regrouping elsewhere must consider contexts and issues beyond climate change as well as concerns and cultures of the local people.

On a different track, Kothari (2014) comments on the politics of climate change discourse while analysing the resettlement policies in the Maldives, where the government proposed to resettle communities of 200 islands into 10–15 safer islands. She shares the local as well as academic concern that climate change is used as a "political tool" by opposing parties, negating "politics of the possible" that could promote more equitable and sustainable alternatives for the future (Kothari 2014: 137). Even with interventions aimed at resilience and adaptation, structural elements of vulnerability (Wisner et al. 1984) can still deprive vulnerable people, or maintaining an unequal status quo could still lead to damages and even casualties (Cannon and Mueller-Mahn 2010; Frerks et al. 2011).

These critical studies do not take a very optimistic view of migration being a solution for problems stemming from climate change, but at the same time negate the view of migration as a desperate act of poor people in fragile, changing or degraded environments. New studies also problematise the option of planned relocation. At different levels, they challenge the framing of climate change as the key or only driver of largely involuntary migration from poorer parts of the world. They also caution against solutions that emerge from such a narrow view, especially when only environmental impacts of climate change are considered, ignoring the socio-economic and political context in which they work. A more positive viewpoint is that people moving out or made to move out of vulnerable places in a planned way are not 'victims' of climate change, but agents of change in an environment of climatic uncertainties and hazard risks (Tacoli 2009; Foresight 2011; Barnett and Webber 2010).

3.4.2 The option to stay back and adapt

While migration can be an effective adaptation strategy, people's response to hazards need not necessarily be moving out, it could also be staying put and adjusting their livelihoods to suit the changed scenario in another form of adaptation. Adaptation can be autonomous or planned, structural or non-structural and in situ or ex situ (Fankhauser et al. 1999; McCarthy et al. 2001). Cooper and Pile (2014) have looked at a broad range of adaptation options. They include making adaptive changes in lifestyles by living in suitable buildings; changing infrastructure land-use or livelihood; migrating; or making changes in the geographic features by building flood defense, seawalls and nourishing beaches.

In the context of climate change, research into intention and behaviour need not necessarily relate to migration, but in situ adaptation options as well. While adaptive actions can aim to preserve the status quo, defending lifestyles and assets, they can also encourage incremental improvements to existing systems, tweaking resource management practices, enhancing livelihoods by income diversification and improving disaster preparedness measures and sustainable development programmes (Huq et al. 2003; Smit and Wendel 2006). Remittances from migrants can supplement household income and contribute to a community's ability to stay back and be productive rather than being unable or unwilling to move out. Such adaptation can also prevent larger-scale migration of whole households or communities in an "unplanned and unpredictable" way, or to vulnerable areas (Foresight 2011: 14).

Climate change adaptation involves planning for huge uncertainties with insufficient data, boosting resilience to a spectrum of shocks and stresses – by way of providing safer structures, infrastructure and services, including emergency warning and response systems (Dodman and Mitlin 2013). The adaptation decision-making process, however, involves looking at conditions under which people make decisions that are beneficial to their future, even with limited data at hand and under conditions of uncertainty (Gowda and Fox 2002; Grothmann and Patt 2003). As Barnett and Webber (2010) sum up: "In many ways migration can also contribute positively to adaptation to climate change, notably through the way it can build financial, social and human capital. There are policy measures that can enhance the contribution migration can make to adaptation. . . . However, migration in response to climate change also has its risks." The following section discusses evidence for such migration and its outcomes.

3.5 Recent evidence for the climate-environment-migration nexus

3.5.1 Mobility outcomes of stresses and shocks

This review looks at how climatic and environmental stresses and shocks relate to migration in the literature. Among stresses, droughts and rainfall variability have been associated with migration. For instance, Findley (1994) found that the 1983–1985 drought in Mali halved cereal and livestock production, making the majority of families depend on migration and remittances. However, the average migration rate did not increase during the drought – possibly due to the existing level of migration already being at a saturation point, or perhaps due to the role of remittances or food relief. However, the migration cycle became shorter, with short-cycle migration more than doubling, and more permanent migration declining. The short-cycle migrants came from families with lower average incomes compared with the long-cycle migrants.

Henry et al. (2004) studied migratory pathways during 1960–1999 in Burkina Faso, using environmental typologies of origins and destinations, based on rainfall variations and land degradation. Environmental factors were seen to influence the probability to migrate and the selection of a destination. It is stresses such as land degradation rather than episodes such as droughts that influenced migration more. A larger proportion of people living in areas of unfavourable conditions and land degradation migrated, when compared with those living in areas with better land, even with unfavourable climatic conditions. In another study, Henry et al. (2004b) note that people from the drier regions are more likely to migrate temporarily or permanently to other villages than those from wetter areas. The study also showed that short-term rainfall deficits led people to migrate to other villages on a long-term basis, but reduced short-term moves to faraway places.

Barrios et al. (2006), however, found that shortages in rainfall increased urbanisation levels in sub-Saharan Africa. Van der Geest's (2011) study on the Dagara people's migration in Ghana showed that rather than degradation and natural disasters, structural differences in agro-ecological conditions played an important role in their movement and that environmental factors act in complex interplay with economic, political, social and cultural ones. Studying the outmigration in Ghana's forest-savannah transition zone using household surveys, Abu et al. (2014) have noted that even in the case of people perceiving considerable environmental stress, climate may not be the primary driver for migration intentions, unlike socio-demographic factors such as age, household size and current migration status.

Studies elsewhere have also shown a mixed influence of climatic and socio-economic variables on migration. Munshi (2003) studied migration between Mexican provinces, using rainfall variation in the migrants' points of origin to identify how networks at their destinations influence migration patterns. The rationale is that low rainfall in the distant-past increases the number of older migrants in the network, contributing to better access to jobs for migrants who come later.

Migrants with better networks not only had better chances to find better-paid non-farming jobs, but also retain jobs compared with others.

Saldaña-Zorrilla and Sandberg (2009) studied migration between Mexican municipalities. Their spatial econometric model showed that municipalities with greater income dips and recurrent disasters (during the 1990s) had higher emigration rates. The results suggest that people may be migrating for better income prospects, a rational economic choice based on estimated future returns, taking into account asset losses, available finances and expected net assets. Kniveton et al. (2008) note that drought in general appears to increase short-term rural-to-rural migration, but does not affect, or even decreases international, long-distance moves.

Turning to shocks, studies have shown that though disasters can lead to large-scale displacement, such displacement is often temporary. People come back to rebuild, and when away, they prefer to live in places that are familiar to them (Lonergan 1998; Black 2001; Castles 2002; Perch-Nielson et al. 2008; Piguet 2008). In an empirical analysis, Drabo and Mbaye (2011) have reported that disasters have an impact at the same time of the event as well as after a time lag, whereas other climatic events – such as rainfall variability – have only a lagged effect.

In the case of storms and floods, mitigation and adaptation capabilities are limited than for events such as extreme temperature or droughts that give people more time to prepare. Halliday's (2008) study in El Salvador shows that following shocks affecting farming, men spent more time in farm work, and more migrant men lived in the US. Most households did not send women for farming locally or wage labour abroad. In their Sub-Sharan Africa study, Marchiori et al. (2012) noted that weather anomalies have a significant and robust impact on average wages that, in turn, international migration. As Black et al. (2013) argue, long-term displacement appears relatively rare in the case of extreme environmental or climatic events, however, displacement figures often note the peak level of movement at the emergency phase rather than longer-term migration. Rapid returns are usually not systematically documented (Black et.al. 2013).

Non-climatic disasters also show similar results. A study of mobility patterns after the 2010 earthquake in Haiti showed that people who left the capital city within three weeks of the event moved to places where they had social networks (Lu et al. 2012). However, the 2001 earthquakes reduced migration of women, but not men, leading to an increase in women's domestic work hours, but not men's (Halliday 2008).

Among large-scale studies, a two-year empirical study called Environmental Change and Forced Migration Scenarios (EACH-FOR) collected evidence from eight detailed cases in South Asia and south-east Asia. It found that households tried to offset losses suffered in climate-related risks such as rainfall variability by migrating seasonally, for short periods or permanently thereby enhancing food and livelihood insecurity. The study looked at the changes of natural and human-made environment among the causes of migration and explored the linkages and consequences at local, regional and national levels (EACH-FOR 2009). As EACH-FOR (2009) shows, migration is common in all eight sites, but almost entirely within state borders, undertaken mostly by men, but with growing participation by women.

Within the EACH-FOR study, findings suggest that migration is largely for better income, and better education and skills, as demonstrated by case studies in Thailand, Vietnam and Peru (Warner et al. 2013). It was towards cities and better farming areas in case studies in Ghana, Bangladesh, Tanzania; nearby cities as in Peru, India; and towards places with mining operations or industries in case studies in Ghana, Thailand and Vietnam (Warner et al. 2012). Overall, the study suggested that rainfall variability has an impact on household income and migration decisions, and income diversification, including by migration, and education enhance resilience and adaptation. Those who lack such options run the risk of being trapped on the "margins of decent existence" (Warner et al. 2012: 5).

The notion of people who do not move, or trapped populations, comes up as one of the key concerns in the Foresight report (2011) as well. Based on 70 evidence papers, focusing on three ecological regions, namely low-elevation coastal areas, drylands and mountain regions, Foresight (2011) noted that environmental change will affect migration through social, political, economic, environmental and demographic drivers through a set of complex interactions. The report argued that people are as likely to migrate into and out of places of environmental hotspots like urban flood plains of Africa and Asia. While migration is primarily for livelihood-related reasons, environmental change will enhance people's exposure to natural hazards, and thereby influence migration patterns. The report called for planned and well-managed migration options to address the issue of immobility, as millions could be 'trapped' in vulnerable areas of poor countries, unable or unwilling to move, or could move into environmentally vulnerable areas. That means better planning of migration destinations, especially cities prone to flooding, water shortages and inadequate housing.

It may be argued that climatic and environmental stresses and shocks influence migration in positive and negative ways and the migration response of people is contextual; and migratory movements vary across time and space. An insight from the literature is that when people have the choice and capacity to migrate, it can help them adapt better to climate change by gaining new and improved livelihood options, adding to resilience through remittances and skill sharing (Foresight 2011). The contribution of Foresight (2011) lies in such a clear framing of migration in the context of climate change, highlighting its multi-causal and complex nature, and at the same time squarely addressing the issue of immobility. Both these issues are very much relevant, especially in the case of Bangladesh. It is a country that is exposed to natural hazards (Kreft et al. 2015; Harmeling 2012), effects of climate change (Huq and Ayers 2008), high levels of rural poverty and urban growth driving migration (Marshall and Rahman 2013) and high population density (BBS 2011).

In another desk study, echoing earlier concerns about environmental change and conflict (Homer-Dixon and Percival 1996; Myers and Kent 1995), Reuveny (2007) argued that people from developing countries might be more likely to migrate from climate-affected areas, potentially causing conflict in receiving areas. Using Centre for Research on the Epidemiology of Disasters (CRED) data, this study calculated total numbers of people affected by natural disasters of a certain type during 1975–2001 and related it with conflicts in the receiving areas. He found that conflict was present in 19 out of 39 episodes of such migration. Reuveny (2008) followed up this research with case studies of Hurricane Katrina of 2005, the US Dust Bowl in the 1930s, and Bangladesh-India migration since the 1950s. In the case of Bangladesh, the majority of the people settled in new places have led to conflicts with the local people of their destinations. In the aftermath of Hurricane Katrina, many people came back home the following year, while others preferred to stay away. Though these migrants were welcomed in general by their host communities there were tensions; the author concludes, "large-scale ecomigration could increase international tensions, perhaps instigating terrorism recruitment" (Reuveny 2008: 10).

Reuveny (2007: 659) argues that mass migrations associated with environmental degradation might trigger conflict in receiving areas because of a range of factors. These factors include "competition" for resources; "ethnic tension" between the migrants and the hosts; "distrust" between sending and receiving communities, pre-existing "fault lines" that mark tensions on account of socio-economic issues, and a set of "auxiliary conditions" such as an underdeveloped economy and civil strife. Reuveny (2007: 660) further claims that though the model applies to "climate change-induced and ordinary migration," the scope and speed of the latter could be far greater in the aftermath of "evermore frequent and intense droughts and storms."

While none of the above social and environmental conditions is necessarily or exclusively associated with climate change or migration, the argument in favour of "climate-change-induced" conflict sounds far-fetched. As Raleigh (2008: 35) critiques this study: "The suppositions and conjectures mask poorly designed models of causation without reference to the mechanisms, opportunities, underlying motivations, past histories, role in international assistance and government policies on migrants." The issue with such proposed causal claims is that they fail to take into account actual migration processes. The bold assumptions are seldom based on robust and detailed models (Findlay 2011). Environmental change-related movements tend to be short distance, with migrants choosing destinations where they have social networks, especially largely cities of the global South, and such movement seldom gives rise to conflicts (Findlay 2011).

3.6 Policy implications

Recent literature suggests that recurrence of climatic and environmental stresses and shocks, including extreme events, makes Bangladesh highly vulnerable to climate change (Findlay and Geddes 2011; Black et al. 2013; Penning-Rowsell et al. 2013), requiring very effective adaption measures (IPCC 2012; Planning Commission 2012). Events and processes such as cyclones, floods, coastal/delta erosion and water shortage could affect migration in 50 years, largely within the country or to nearby countries (World Bank 2011; Foresight 2011). In this context, migration plays an important role in enhancing the adaptive capacity of people, as explained in chapter 3.

The benefits of migration, however, are often debated. On the one hand it is seen as an activity that contributes to better resources, aiding better income distribution with remittances from the migrants (Spaan et al. 2005); and on the other hand it is dubbed as a mark of economic dependency, hindering local development (Heremele 1997). Alternative readings, such as the 'time perspective' (Rahman 2009), take a more balanced view of migration with its negative costs dominating in the short term, but eventually spurring development. The policy analysis in chapter 8 is built around the idea that migration is part of human economic activity, and it can help people escape as well as offset losses and recover from climate- and environment-related stresses and shocks, especially in the context of climate change. It is the structural context of a country – in terms of investment environment, institutions, residency, education and employment rights – that often determines the overall benefit for migrants (de Haas 2012).

However, policies tend to take a negative view of climate change-related migration (Barnett and Webber 2010), with National Adaptation Programmes of Action (NAPAs), the basic plans for adaptation in Least Developed Countries (LDCs), often calling it a failure of adaptation problem (Sward and Codjoe 2012). Such a negative view can badly influence policy formulation and implementation, limit options for migration and resettlement and seriously restrict the benefits of migration (Laczko and Aghazarm 2009), in effect trapping people in environmentally vulnerable places (Foresight 2011). Such policies tend to fail too (de Haas 2006; World Bank 2010). Instead, when allowed, planned and facilitated migration can be an effective form of adaptation to climate change, as the literature shows. It helps people adjust to their environment exposed or degraded by climatic stimuli, by minimising harms, allowing alternative livelihood opportunities in line with social norms and processes (McCarthy et al. 2001; Adger et al. 2005; IPCC 2012). While the international community seeks more action and cooperation on adaptation (UNFCCC 2011), evidence shows migration can offset the impacts of environmental shocks and stresses (McLeman and Smit 2006; Barnett and Webber 2010; Tacoli 2009; Foresight 2011; ADB 2012). It can reduce vulnerabilities and enhance households' adaptive capacity and allow them to gain better access to natural resources, livelihoods, social networks and markets (Gerlitz et al. 2014).

Migration can involve a broad spectrum of activities – including escaping risk (Adger et al. 2005; IPCC 2012; Penning-Rowsell et al. 2013), surviving extreme events (Findlay and Geddes 2011; Black et al. 2013), diversifying climate-affected livelihoods (Tacoli 2011) and so on. Migrants' remittances boost adaptive capacity back home (Guzman et al. 2009; Warner et al. 2009; Foresight 2011). Migration can be a reasoned response embedded in existing livelihood patterns (Gardner 2009). It can be undertaken for short or long durations to nearby villages or faraway cities or even overseas. Even when influenced by climate-related vulnerabilities (Banerjee et al. 2012), migration is largely driven by socio-economic and other factors making up one of many complementary livelihood choices (Kniveton et al. 2008). Environmental change influences drivers of migration across a range of overlapping social, political, economic, environmental and demographic spheres (Foresight 2011; Black et al. 2011). Cutting across these spheres, disasters, development projects, environmental degradation, shortages, poverty and market changes can often act together to amplify vulnerabilities (Stal and Warner 2009).

3.7 Conclusion

Recent research focusing on the linkages among climate, environment and migration has generated new methods and evidence, and added new dimensions to the field of migration studies. While climate change poses risk to human security, mainly by affecting livelihoods, research has shifted focus away from generalised notions of climate change and apocalyptic imageries involving mass migrations of impoverished people leading to shortages and conflicts. Nonetheless, such neo-Malthusian notions still gain currency in research and policymaking.

Slow-onset climate processes as well as rapid and extreme weather events and changes in the environment have different and sometimes diametrically opposite influences on migration. Such phenomena can make livelihood stresses even worse, forcing people to migrate in search of better income. They can also trap people in their localities for a variety of reasons – to take care of their farm, by depriving them of the basic resources necessary to migrate or by making them move to vulnerable places (Foresight 2011). As research continues, there is a need to take a closer look at specific geographies, especially areas exposed to extreme and uncertain weather, such as low-elevation coastal areas, drylands and mountain regions (Foresight 2011).

At the same time, along with better evidence for the multi-causality of migration, behavioural and cognitive approaches to climate- and environment-related migration pushes the boundaries of migration studies, giving new dimensions to classical migration theories. New research takes into account multiple influences upon the elaborate decision-making process involved in it. It looks at individual, household, environmental and structural dimensions of the drivers of migration (Kniveton et al. 2011). Such detailed understanding of a multi-faceted and complex process has policy relevance in the context of environmental change. As climate and environments change and weather extremes and uncertainties place livelihoods and human security at risk, this research can help give more insights and options of interventions that lead to better adaptation and resilience of local communities. In this context, it is important to gather evidence from different settings, especially areas that are vulnerable to natural hazards and regarding different forms of migration, such as Bangladesh.

Notes

1 Carrington, D. (2016). Climate change will stir 'unimaginable' refugee crisis, says military *The Guardian*, December 1 [online] www.theguardian.com/environment/2016/dec/01/climate-change-trigger-unimaginable-refugee-crisis-senior-military accessed on 30 March 2017

2 Jeffrey, S., and Rehman, A. (2017). Desperate exodus of the climate refugees, Letters, *The Guardian*, January 9 [online] www.theguardian.com/environment/2017/jan/09/desperate-exodus-of-the-climate-refugees accessed on 30 March 2017

3 Baker, A. (2015). How climate change is behind the surge of migrants to Europe, *Time*, September 7 [online] http://time.com/4024210/climate-change-migrants/ accessed on 30 March 2017

4 Ferrie, J. (2017). Climate change and mass migration: a growing threat to global security, *IRIN News*, January 19 [online] www.irinnews.org/analysis/2017/01/19/climate-change-and-mass-migration-growing-threat-global-security accessed on 30 March 2017

4 Stay or move?

Migration decisions amidst uncertainties

At one level, this chapter looks at Bangladeshi villagers making livelihood choices amidst new economic opportunities and development patterns. At another level, it looks at how climate- and environment-related stresses and shocks make livelihoods difficult and make additional or better income necessary. It builds on peer-reviewed research that analyses focus groups and interviews in 14 villages across three districts affected by multiple hazards and disasters, including floods, riverbank erosion, cyclone and drought. With data, insights and contextual explanation, the chapter argues that it is not only cost-benefit or risk-resilience considerations that matter in livelihood decision-making, but also the ways in which people perceive changes and response options and act according to the socio-cultural acceptance of the choices before them.[1]

> "I worked as a day labourer in the village then. However, most of the time I had no work to do. In 1989 I went to Chalna (a sub-district town in Khulna district) in search of work with a group of seven or eight persons. I got the work of paddy harvesting. I worked for 13 days and earned BDT 1500 (GBP 15.5) and came back home. After two months, I went to Faridpur with the same group. At that time, we got the work of digging of pond. We worked the whole *Kartik* (mid-October to November) month there. Later we went to Jaldumur in Khulna for paddy harvesting in the months of *Agrahayon* and *Poush* (mid-December to mid-February). Thus, I have continued my livelihood as internal migrant and worked in the nearby districts.
>
> "When my elder son left school at 14, I took him with me for work in Khulna and Faridpur. My second son is also a migrant worker. He works in brickfield in other districts such as Magura, Comilla. In 2010, a local contractor asked me [for permission] to hire my son for six months in return for BDT 35,000. When I agreed, he gave me BDT 20,000 in advance. In 2011 my son also went to work in a brickfield in Comilla district."
>
> – Muhammad Kazim (name changed to protect identity)

4.1 Moving from the margins

Kazim, one of the respondents in our qualitative research, told us this story in Gabura. He is 39 and lived in a small, dilapidated house that leaked when it rained.

1 This chapter draws substantially in research and content from Martin et al (2014).

At that time, his village looked barren and muddy. On rainy days, it looked like a flat disc floating on the water, thick clouds hovering above, pouring it out. Gabura has always been vulnerable to cyclones, and it became one of the worst affected villages during Cyclone Aila, grabbing international headlines.

Kazim said the cyclone pushed him even deeper into poverty as there was no way people could cultivate anything in the village after the storm surge. It was sea salt everywhere – in ponds, soil, surface water and aquifers. Nothing grew there. Kazim's migration forays became longer and more frequent. He said he pursed many livelihood activities – cattle-herding, casual labour, rice harvesting, digging, brick making. Somehow, he had to make both ends meet.

Kazim's father, like most people of his generation here, was not so much of a migrant. "My father is a farmer. He also works as a labourer in other (people's) farms." The cyclone apart, Kazim finds several climatic and environmental features making livelihoods precarious in Gabura – uncertain rains, floods, salinity of the soil and water, which seems to be on the rise over the past few years. Riverbank erosion is a threat too, but on the scale of hazards, that plays a minor role here. People have many other factors to worry about in this village. By 2015, five seasons of rains have washed away much of the sea salt left by the cyclone in the soil and water of the village, and filled the village ponds. Gabura became a green, fertile place once again. People still migrated from Gabura – but they said they were not desperate as they were during the years following the cyclone. They still knew Gabura is intrinsically a vulnerable place, as their narratives in this chapter shows. Still the sentiment – as always – appeared to be on the lines that "this is our land; our ancestors' lived and died here; we love it; we would not like to leave."

Statistics show that there is outmigration from coastal areas of Bangladesh, as the next chapter shows. Still our qualitative study revealed the seasonal, temporary nature of much of the migration, as Kazim's story shows. Migration appeared to be driven by poverty and for better income amidst widespread poverty in villages and economic opportunities, especially in the informal economic sector – in villages, towns and cities. Climate- and environment-related stresses and shocks often work in the background, making livelihoods difficult and uncertain, if not precarious. That means migration in the context of climatic and environmental stresses often involves mixed motives. Our research tried to tease out these multiple strands of migration narratives.

4.2 Why do people move?

Qualitative research considered three research questions:

1 What are the various climatic and environmental stresses and shocks that affect rural areas of Bangladesh? (We sought local people's perspectives on these questions, following up on the literature discussed in chapter 3.).
2 What are the changes and uncertainties in climate and environment that people experience in their locality; how do they perceive the risk they pose?
3 To what extent do people acknowledge the role of these experiences and risk concerns in their migration decisions?

We chose three geographically distinct settings for our qualitative research – a drought-prone district, a flood and erosion-prone district and a third district exposed to frequent cyclones. Within these districts, we chose 14 villages, as explained below. To set the stage for our enquiry, we did a preliminary survey of each of our 14 study villages exposed in different ways and degrees to climatic and environmental hazards. Then we held focus group discussions in each of these villages. While understanding the local people's take on these hazards, we were curious about how these factors influenced their decision to stay in their place or move out – over short or long durations to places that are close by or far away, with their families or alone. Therefore, we conducted semi-structured interviews with 20 respondents – looking more deeply into aspects discussed in focus groups.

The key challenge in research probing climatic and environmental factors that influence migration is to understand their complex linkages with livelihoods and, therefore, people's decision to stay or move out of their villages for short or long durations. Bangladesh villages are exposed to a wide range of climatic and environmental stresses and shocks. Therefore, we chose broadly three parts of Bangladesh that are exposed to these stresses and shocks. In the south, close to the Bay of Bengal coast, villages are exposed to cyclones and the storm surges they cause. Much of the country is exposed to floods and riverbank erosion. In the north-western part of the country, drought and water shortage are a problem. We selected three districts of Bangladesh based on how they are exposed to different climate- and environment-related stresses and shocks – Nawabganj in the north-west region of Bangladesh under Rajshahi division, Munshiganj, a part of Dhaka division close to the central part of the country, and Satkhira, belonging to the Khulna division in the south (please see table 4.1).

While Nawabganj is exposed to droughts and floods (Habiba et al. 2011), Mushiganj witnesses frequent floods and riverbank erosion (DMB 2010), and Satkhira suffered the impacts of two cyclones, Aila (2009) and Sidr (2006) (DMB 2010; Sultana and Mallick 2015). From these districts, we selected 14 villages based on their level of exposure to these phenomena. They may not be called a statistically representative sample, but a set of examples to show how climatic and environmental stresses and shocks play a role in people's livelihoods and their migration decisions.

In Nawabganj, two *upazilas* (sub-districts) were selected for fieldwork – Nachole and Shibganj. Nachole is part of *barine* land or natural highlands that extend to several villages in the neighbouring districts of Naogaon, Rajshahi and elsewhere. Being part of *barine* land, drought is a chronic problem in villages here, with the groundwater level declining to 150 feet below the ground level in some places, as local villagers narrated. That is double the depth of the wells they had two decades ago. From mid-November to mid-June, often water is scarce. In Mohnohil, one of the study villages in Nachole, villagers said that drought affects rice production, along with high input costs and low yield, leaving farmers

Table 4.1 Field study villages and climatic and environmental hazards

District	*Sub-district*	*Village*	*Hazard exposure*
Munshiganj	Lohajong	Kolma	Riverbank erosion, flood
		Mandra	Flood
	Sreenagar	Bhagyakul	Riverbank erosion, flood
	Sirajdikhan	Charipara	Flood
		Char Sonarkanda	Flood
		Dakerhati	Flood
Nawabganj	Nachole	Mohanohil	Drought
		Khesba	Drought
	Shibganj	Chorpka	Riverbank erosion, flood, drought
		Durlogpur	Riverbank erosion, flood, drought
Satkhira	Shyamnagar	Gabura	Cyclone, riverbank erosion, salinity
		Khailasbonia	Cyclone, riverbank erosion, salinity
		Paddapukur	Cyclone, riverbank erosion, salinity
		Khutikata	Cyclone, riverbank erosion, salinity

burdened with debt. In Khesba, another study village in Nachole, there is no work during the dry season and about 10 per cent of the work force migrate to cities such as Chittagong, Sylhet, Dhaka and Narayanganj. They work as masons, helpers to masons or pull rickshaws and vans, three-wheeled carriages that carry people and goods.

The other field site in Nawabganj district, Shibganj, is located on the river Padma. Two villages from a *char* area were selected for fieldwork – Durlogpur and Chorpka. These are low-lying areas formed of silt and surrounded by water most of the year, prone to flood, erosion and, ironically, drought. The villages are located along Bangladesh's border with India on the north-west of the country, 20 km away from Shibganj *upazila* headquarters, reachable only by a two-hour motorboat ride. Over the past 40 years, riverbank erosion has led to loss of homesteads and farms. Three months of monsoon rains leave the villages flooded, making it impossible to cultivate in Durlogpur. Chorpka is a stretch of land that emerged in 1970 over the river Padma, people began inhabiting it as Bangladesh started nation-building in its first decade of independence in 1971. Still most of the arable land (about 80 per cent as the villagers estimate) here goes under water during monsoon season.

In Munshiganj, three *upazilas* were selected for fieldwork – Lohajong, Sreenagar and Sirajdikhan. These low-lying villages here are also vulnerable to regular flooding and riverbank erosion. Sreenagar and Lohajong are on the banks of the river Padma; Sirajdikhan sub-district is located by the river Dholeshwari. Bhagyakul village in Sreenagar, located north of the river Padma – about 50 km from Dhaka – faces severe riverbank erosion. The villagers said only 2000 square feet of land around the old fish market is now left intact after the village lost much of its original area of 300 hectares to the river. The displaced villagers have taken shelter in the neighbouring villages of Mandra, Charipara, Rarikhal and Khamargaon. Some of them bought land, some rented property and some squatted on government and

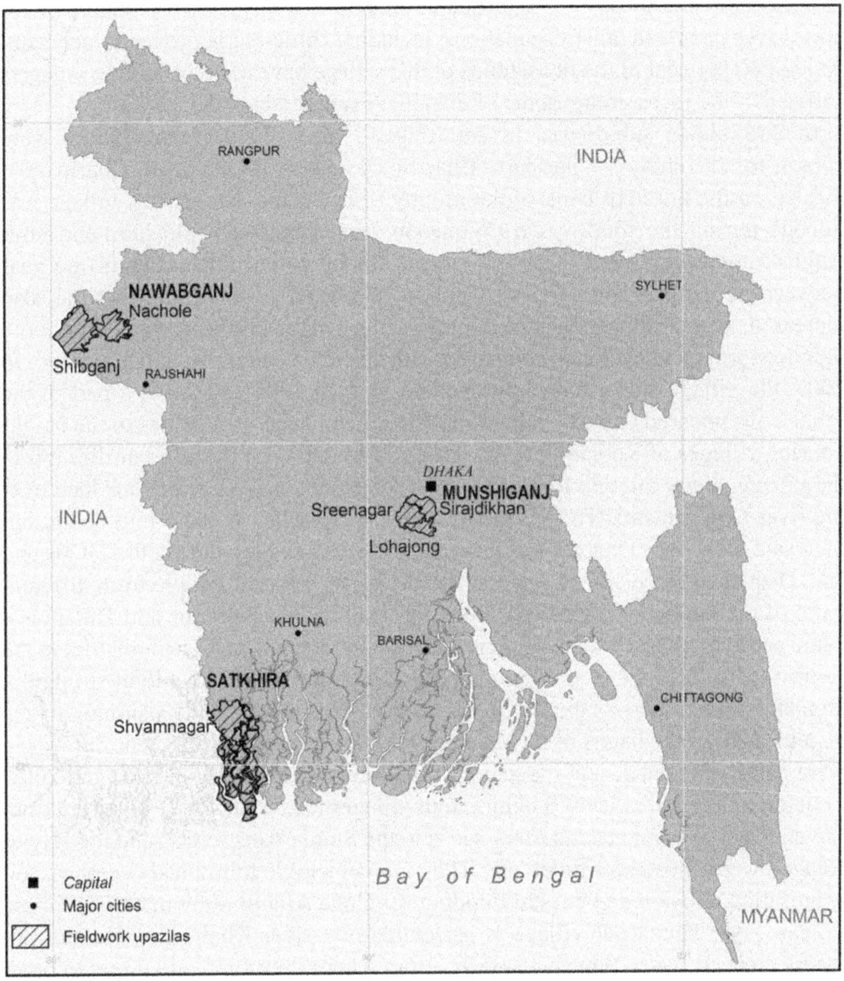

Figure 4.1 Map of Bangladesh showing the sub-districts chosen for the qualitative study[2]

private land. As if to compensate for the losses, the local villagers cultivate their farms up to three times a year, they said in our focus groups' discussions.

From Lohajong, two flood-prone villages were chosen. One of them, Kolma, located 40 km away from the capital city Dhaka, is an island on the river Dholeshwari. Over the past 15 years, migration has gained prominence in this traditional farming village. During non-farming season, 15–20 per cent of the work force migrate to other districts, such as Dhaka, Gazipur and Narayangang to engage in casual labour in construction sites or pull rickshaws. About five per cent of the households have someone working abroad. The other field site in Lohajong sub-district is Mandra.

2 Map drawn by Pedram Rowhani

Until about 30 years before, the principal occupation of people of this village was farming and fishing. Loss of land due to erosion and increase of production cost now drives people to other occupations, including cattle-rearing (over 30 per cent). At least 40 per cent of the households of this village have migrants, as the villagers narrated in the focus group; some of them have migrated abroad.

In Sirajdikhan sub-district of Munshiganj, three flood-prone villages were chosen for the study – Charipara, Char Sonakanda and Dakerhati. Charipara is located on the northern bank of the mighty river Padma. Earlier this village was flooded during the southwestern monsoon season between mid-June and mid-August, however, over the past 10 years, floods come much later in the year between September and early November. The water level of the river has also increased, as the villagers narrated. Of late, the village has been witnessing more thunderstorms and cyclones compared with earlier decades, the villagers said. In 2007, the village was affected by cyclone Sidr. In 2006 and 2007, a part of the village disappeared due to riverbank erosion, displacing 40 families to the neighbouring villages of Kamargaon, Rarikhal and Mandra. Of the 100 families left in the village, many migrate for work. Chor Sonakanda is a piece of *char* land over the river Dholeshwari. The villagers cannot recall when it came into being, but they said close to 30 per cent of its population settled here during the 1970s and 80s. Displaced by riverbank erosion and driven by poverty, people from different parts of the country – especially Rangpur, Gaibandha, Faridpur and Dinajpur – came and settled here. Many of them came first to Keraniganj, a sub-district in the southwest of Dhaka – a guerilla operations hub during the liberation struggle – and later moved here, as the villagers narrated. This village and Dakerhati, that is also located on the banks of the Dholeshwari, are exposed to frequent floods and were badly affected during the major flood events of 1988, 1998, 2004 and 2007.

In the southern district of Satkhira, four villages located about 15 km away from Shyamnagar *upazila* headquarters, close to the Sundarbans forests and the Bay of Bengal, were chosen for fieldwork. They are vulnerable to tropical cyclones, soil salinisation, erosion and coastal flooding. Cyclone Aila of 2009 devastated these villages. Gar Kumarpur village is surrounded by river Khalpetua, that the villagers cross to reach Nowabeki Bazar, from where they shop and move to other parts of the district. The local people make a living by farming, fishing, collecting shrimp fingerlings and gathering honey, wood and leaves from the mangroves. Many are migrant labourers who work in rice farms, brickfields and other fields of work in Satkhira city as well as Khulna, Gopalganj and Jessore districts. Irregular and insufficient rainfall – especially during *Ashar* and *Srabon* – and occasional extreme rain events affect farming and shrimp cultivation, the villagers said.

At Khutikata, a quarter of the people worked outside the village. Combined with those engaged in seasonal and temporary work outside the village, the share of migrants in the village can be as high as 40 per cent. Until shrimp cultivation gained popularity in 1992, a single crop village survived on rice harvest, called the *Aman* paddy. People topped up their income by collecting honey from the Sundarbans during mid-April to mid-May, small shrimp from mid-May to mid-August, fish from mid-September to mid-November and *golpata* (leaves of the mangrove palm *Nipa fruticans*, used to thatch huts and to make bags and mats) round the year.

Gabura village is surrounded by the rivers Kobadak and Khalpetua. The villagers need to cross the Khalpetua for shopping from Nildumur Bazar and Gavura Bazar. Located on the south of the village is the Sundarbans and beyond that the Bay of Bengal, and on its western side, beyond the forest, is the Bangladesh-India border. Many people from here work as seasonal migrants mainly in the rice fields of nearby districts Khulna, Jessore, Gopalgonj or in the brickfields of Dhaka and Munshigonj. The villagers said that shrimp cultivation reduced work opportunities in the village, as its maintenance requires just one or two people, but a 33 decimal (0.33 acre) agricultural land can accommodate five or six people. Many villagers depend on mangroves; still government restrictions and the risk of attack by robbers and tigers make that option difficult too, the villagers said.

4.3 Approach to qualitative study

The research involved collection of three sets of data. First, village surveys elicited basic information, including geographical characteristics of the area, population, livelihood, educational institutions, health care facilities, types of environmental hazards and crop patterns. The surveys elicited information on land availability, transport facilities, farming, fishing and employment opportunities. This information has been used as the background to understand the socio-economic profile of the village as well as its climatic and environmental features and hazard exposure. Then focus group discussions explored how local climate, environment and livelihoods have changed over the past 30 years. The discussions also covered migration trends, as well as current and future concerns of the local people in terms of climate, safety and livelihoods. The focus groups comprised community leaders and elders. The questions about climatic stresses and shocks included flood, drought, riverbank erosion, salinity of the soil and cyclones. They elicited local responses to the projected impacts of climate change reported, especially, in recent synthesis reports (Walsham 2010; Foresight 2011; Adams et al. 2011a, for instance).

Finally, 20 in-depth interviews probed further how people respond to these changes and uncertainties as well as livelihood challenges and opportunities. Individual in-depth interviews elicited detailed information on migrants as well as non-migrants and their households, and how their livelihoods changed over the last 30 years. The focus was on factors that drove people to migrate and to the extent to which climatic stresses and shocks influenced the migration patterns. The interviews probed cognitive aspects of the migration decision-making. They included questions on the problems the interviewees faced, the potential of migration to solve them, the perceived severity of climatic stresses and shocks and the perceived effectiveness of responses, including migration. They also covered family and social attitudes to migration, social networks that facilitated movement, locus of control, risk-taking and trust in advice given by family, friends and officials as well as private organisations. These observations were compared with the literature based on observed data on climate, environment and hazards. As the paragraphs below show, the climate, environment and migration literature discussed in chapter 2, as the respondent's statements illustrate, elaborate or contradict earlier findings in this field. It is important, as each setting of climatic and environmental migration varies.

In Bangladesh, migration decisions are made in the broader socio-economic context of a transition from rural farming-based livelihoods to a more mixed pattern that also involves different forms of labour and trade. This happens amidst fast-paced city-based economic growth, as discussed in chapter 3, and ensuing large-scale rural-urban movements (Muzzini and Aparicio 2013; Marshall and Rahman 2013); and at the same time exposure to climatic and environmental hazards in a changing climate (Adams et al. 2011a). The argument here is that it is not only cost–benefit (Massey et al. 1993) or risk-resilience (Wisner et al. 2004) considerations that matter in livelihood decision-making, but also the ways in which people perceive changes and response options and act according to the socio-cultural acceptance of choices before them.

Choosing one of these options involves a decision-making process. Theory suggests that migration could involve a household-level decision to improve income, a quest for social mobility, search for a better place to live and a perceived linkage between migration behaviour and rewards in a new location (De Jong and Fawcett 1981). Even when migrating from a village to another village, a town or a city, this may not always lead to formal jobs or comparatively better living conditions, but it still helps the migrants deal with temporary disruptions or reduction in earning back home (Stark and Levhari 1982). Migration decision-making is unique from person to person. Individual values and attitudes (Ritchey 1976), feelings and an exercise of independent will or agency (Stark and Bloom 1985) play a role in it. At the same time, social structure of communities and individual migrants' characteristics and status within the structure also have an effect on migration (Ritchey 1976).

While this enquiry falls within the realm of behavioural research, this chapter pushes the envelope to include socio-cognitive variables that influence people's motivation for migration and their decision-making patterns under uncertainties (Grothmann and Patt 2005). While cognitive research tends to focus on individual-level decision-making, it could also explain how human systems, such as households, communities, cities or even developing countries adapt to climate change (Grothmann and Patt 2003).

The cognitive enquiries, however, tend to assume that people analytically assess and calculate the desirability and likelihood of possible outcomes, thereby underplaying or ignoring feelings about objects, ideas, choices, mental images and emotions. That is because theories of choice under risk or uncertainty tend to be cognitive and consequentialist, using rational choice models (Leiserowitz 2006). A closer look reveals the subjective nature of climate experience, as climate is closely connected to the identity of people in a particular place. It is linked to the emotional, symbolic, spiritual and environmental values they attach to the place and people's identities. Identity refers at the same time to social categories of an individual and to the sources of his or her self-respect or dignity (Fearon 1999). Retaining and reinforcing one's identity is a powerful driver for decision-making. Often it is closely linked to one's place (hence the term 'place identity') and its climate and environment.

Perceptions of climate risks and response options are often shaped by personal observations of a changing environment, belief in God and significance of home and responses (Mortreux and Barnett 2009). That is why cultural products such as folk music, poetry, theatre, worship and festivals often reflect features of local climate. For instance, Christmas roughly coincides with winter solstice in the northern hemisphere, festivities, gifts and candles brightening some of the shortest and gloomiest days of the year in many European countries. As Hulme (2008: 7) argues, "registers of climate can be read in memory, behaviour, text and identity as much as they can be measured through meteorology." It follows that in the face of climate and environmental change, decision-making is not a fully rational, deliberative, analytical act, but rather an "emotionally driven experiential system" (Epstein 1994).

This route of such culture-based enquiry is a road less travelled in climate-related migration research. Still, on a related but different track, Kuruppu and Liverman (2011) built on Grothmann and Patt's (2005) model, adding affective heuristics, intention, implementation plan, development and the stages of change. Heuristics are mental toolkits for decision-making and problem solving, and an affective heuristic involves emotions influencing the cognitive processes. In their study in Kiribati, Kuruppu and Liverman (2011) noted that people wanted to adapt better with more effective water management practices when they perceived climate change as a process that they could feel and relate to; and the more the people believed in their own capabilities, the more they wanted to take up such measures.

On the other side, emerging research also shows that cultural practices could also play a negative role, as not only resource constraints and socio-economic factors limit adaptation choices, but also psychological factors, habits and perceptions of climate variability (Dang et al. 2014). Migration scholars note that barriers to movement could be internal as well as external – people may not be able to move or in some situations, they just do not want to move despite the risks involved in staying at a vulnerable place. Black and Collyer (2014: 52) argue that distinguishing between these two scenarios could be extremely difficult, requiring a "nuanced reframing of migration theory" concerning "migratory space, local assets and cumulative causation."

To explain this decision-making process, this chapter has been structured on the model of migration (Figure 4.2) based on earlier work dealing with adaptation decision-making in the context of climate change (for instance, Grothmann and Patt 2005; Kniveton et al. 2008 and Kniveton et al. 2011). The framework comprises environmental beliefs and a set of behavioural components. The environmental beliefs comprise the respondents' narratives of environmental stresses and shocks based on their own experience. The behavioural components comprise three factors – people's experiences of changes and uncertainties in their locality, perceptions of the risks they pose and the way people make migration decisions informed by these experiences, perceptions and a set of socio-cultural and cognitive determinants (Martin et al. 2014).

This process provided the data for qualitative analysis. That data has been analysed using the behaviour framework (Figure 4.2) explained below. As the next

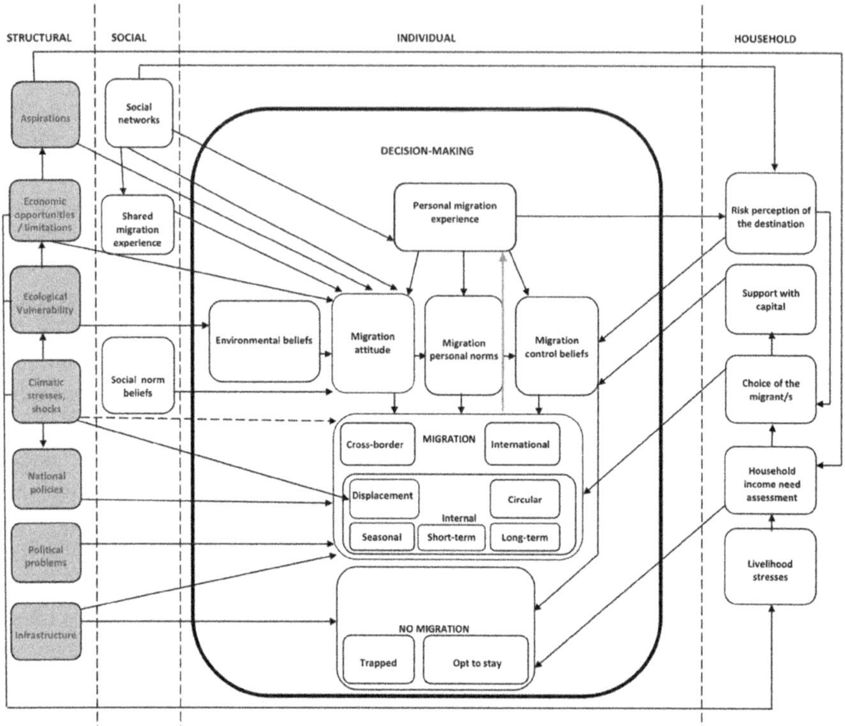

Figure 4.2 Model of migration decision-making

step, the qualitative study probed to what extent migration is a socially acceptable behaviour shaped by experiences of hazardous events and perceptions of future risk. Based on a theoretical model explained below, this chapter explains how the decision to migrate is mediated by a set of 'behavioural factors' that assesses the efficacy of different responses, their socio-cultural acceptance and the ability to respond successfully.

4.4 Modelling individual decision-making process

When you read tabloid newspapers – or even many broadsheets – you get these vivid images of climate change. Rising seas, fierce storms, food shortage, conflict and people fleeing. Very dramatic, to the point of being apocalyptic. While climate change does involve dramatic moments – storm surges, falling ice shelfs for instance – and large-scale forced migration, the connection is not so simple. It is not always the case of people suddenly finding their place being flooded and deciding to pack up and go. These sudden interruptions and movements happen, especially in the case of riverbank erosion in Bangladesh or flash floods. However, more often than that, climate-related migration is a slower and less dramatic process.

There is a decision-making process involved in people staying or leaving a place. It is all about human agency – people making the best out of their situations, limitations and opportunities. The problem with the apocalyptic visions is that they project an image of utter helplessness of what is called 'climate change victims.' My argument is that people affected by climate-related hazards are survivors, not victims. Therefore, studying climate-related migration is all about understanding how people behave under trying circumstances.

To better understand people's narratives about influences over their migration decision-making, we can broadly classify them into a set of environmental and behavioural factors. The qualitative part of our study has been influenced by a model that a few colleagues developed at Sussex. One of the influential models is called MARC (figure 4.2), that means Model of Migration Adaptation to Rainfall Change (MARC), proposed by my colleague Dr Christopher Smith (Smith et al. 2010) and tweaked further in a paper led by Prof Dominic Kniveton. The MARC model identifies risk perception and perceived adaptive capacity as key factors that influence migration decision-making (Kniveton et al. 2011). Kniveton (Dom, as we call him), a climate modeller with a background in atmospheric physics, supervised the doctoral work of Chris and several others like me who researched climate, hazards and migration at Sussex. His teammate and our co-supervisor was Prof Richard Black, a human geographer, famous for his critique of the 'climate refugee' discourse, as narrated in the previous chapter. He led the Foresight (2011) report on environmental change and migration. As I was finishing my doctoral studies, he moved on to SOAS.

Chris has a background in informatics and he built a computerised model called an Agent Based Model (ABM). MARC is an ABM that can be used to simulate behavioral responses of individuals and households to climate signals and interactions between different social actors (Kniveton et al. 2008). The idea is to identify or hypothesize the rules of behavior for migration decisions amidst multiple stimuli (Piguet 2010).

The model incorporates related components such as social and environmental inputs that shape individuals' decision to migrate at four levels – structural, institutional, individual and household. Practically it works like a video game, illustrating movement of people under different scenarios in a simple, animated, graphic display. It showed how people migrated in Burkina Faso as a response to uncertain and scarce rainfall and drought. Kniveton et al. (2011) drew from an influential work that probes the decision-making process involved in climate change adaptation (Grothmann and Patt 2005) that analysed two case studies from urban Germany and rural Zimbabwe to explain the cognitive influence of adaptive action. Their Model of Private Proactive Adaptation to Climate Change (MPPACC) separates out the psychological steps involved in taking action in response to perception of changes in the environment, and examines factors that aid or restrict how people act to adapt to changes in their climate and environment (figure 4.3). The model takes into account risk perception and perceived adaptive capacity, largely neglected in earlier research. MPPACC explains why some people show adaptive behavior while others do not (Grothmann and Patt 2005).

As the MARC model (Kniveton et al. 2011) explains, the individual decision-making processes are influenced by household and social characteristics. Social norms – for instance, the acceptance of a man's decision to be away from home for a few months every year, working in a different place – and beliefs enable and limit the influence of broader structural factors. As it was noted earlier, the decision to migrate could mean different scales of movement across time and space; and not to migrate could mean choosing to do so, or being unable to move, or being 'trapped' in a vulnerable setting (Foresight 2011). Structural factors – that include aspects of environment, political economy and policy setting – form the backdrop in which migration decisions are taken.

These factors were considered while gathering people's narratives about their experiences of climate and environment and the perceptions of risk they pose. The MARC model has been modified on the basis of fieldwork findings, taking into account characteristics of climate- and environment-related migration in Bangladesh. It is a step forward from the MARC, as it incorporates migration across different distances, distinguishes between livelihood stresses as well as household income needs, and deals more specifically with aspects of immobility due to people's choice or inability.

The factors considered in my model (figure 4.1) are described below:

1 Environmental beliefs: perceived probability and severity of threats posed by the impacts of climatic stresses and shocks and environmental change.
2 Behavioural factors:

   a Migration attitudes: assessment of migration options and their efficacy as a livelihood choice
   b Migration control beliefs: generalised expectations about the extent to which people think they can control events that affect them (locus of control, as explained in Rotter 1966; Price and Leviston 2016) and the perceived resources available for migration
   c Migration personal norms: self-concept as opinion leader, perceived level of risk and innovation; and
   d Social norm beliefs: perceived trust in and influences of sources of advice, traditions and cultural factors

4.5 Environmental beliefs: Experiences of climatic and environmental stresses and shocks, and risk perceptions

In Bangladesh, there is an open discourse, backed by scientific studies, about climate change impacts. Scientists in the government system, academia and international organisations have published a body of literature about these impacts and have contributed to this discourse. Numerous outreach programmes through NGOs, media and community meetings have spread knowledge about climate change impacts amongst people. Observations of climate impacts from scientific literature have become reference points for framing focus group discussions and interviews about these impacts and people's perceptions of them in our focus groups.

As for uncertainties and variability in rainfall, rain gauge observations show an increase in March–May rainfall by 3.4 per cent and a decrease in June–August rainfall by 1.7 per cent between 1960 and 2003 (Karmalkar et al., n.d.). Climate models predict a wetter future for the country as well as upstream areas of its great rivers, especially the Himalayas, where they originate (Adams et al. 2011a; Dasgupta et al. 2011; Immerzeel et al. 2013). It can lead to an increase in river run-off, leading to flooding and riverbank erosion (Laczko and Aghazarm 2009; Adams et al. 2011a, EM-DAT 2012a).

At the same time, the number of hot days and nights with mean minimum and maximum temperatures have been increasing by 0.15 and 0.11 degree C per decade, respectively, from 1960 to 2008 (Shahid et al. 2012). This rise compares with the IPCC (2007a) finding that the mean annual global surface air temperature has increased by about 0.74 degrees C over the past century, land surface air temperature increasing more than sea surface temperature. Though the average rise appears to be small, the maximum (in November at a rate of 2.7 0C per 100 year) and minimum temperatures (in February at a rate of 3.4 0C per 100 year) have considerably increased (Islam 2013). This increase has been greater during 1978–2007 compared with 1948–1977. Extreme temperature has claimed 700 lives in 2002, 153 in 2003 and 135 in 2009 (EM-DAT 2015b).

The literature also shows that droughts, especially in the north-western region (Warner and Afifi 2014), and other climatic environmental shocks and stresses undermine livelihoods (Gray and Mueller 2012; Findlay and Geddes 2011; Poncelet 2007). Not necessarily related to drought, water stress due to a variety of reasons is another environmental stress that people face. The literature shows that the reasons include reduced inflow of fresh water into river systems, the over-extraction of groundwater, drainage congestion and river flow restriction caused by India's Farakka barrage across 16.5 km from the international border at Nawabganj district. These impacts appear to cause a northward movement of the salinity line and degradation of mangrove forests (WARPO 2006; Swain 1996). Saline water left by cyclonic storm surges is another reason for the water stress in coastal villages. This has arguably led to salinisation of fresh water resources and soil, making agriculture impossible for many months after the events. The projected increase in the intensity of tropical cyclones and consequences (Peduzzi et al. 2012) could lead to more coastal areas coming under the impact of storm surges in the future (Adams et al. 2011a).

As a result of this combination of factors, Bangladesh villages are exposed to gradual onset climate-related stresses and sudden shocks, including water shortage, cyclone, floods and coastal/delta erosion (MOEF 2005). The focus groups and interviews suggest that people across three geographic zones of Bangladesh – coastal, midland plains and northeastern highlands – share certain common concerns. One such concern reported is irregular or uncertain rainfall that affects farming. Focus groups in two villages each in Nawabganj and Satkhira and three villages in Munshiganj mentioned such rain as something that "affects," "damages" or "threatens" farming. The perception in general is that rainfall is insufficient for agriculture, even in places that are threatened by occasional floods. The respondents used the terms "unpredictable," "irregular" and "decreased" to describe the rainfall trends.

Flood and riverbank erosion are two other common problems that the respondents have reported from all three regions. In Munshiganj and Nawabganj, the villagers consider the 1998 flood as the worst calamity in recent years. People in all three regions felt that there could be more crop failure in the coming decades due to floods. At Bhagyakul in Munshiganj district, many respondents said that riverbank erosion has made floods even more fearsome. For example, local people noted that since 1975–1976, the River Padma has eroded its banks, widening to two miles, submerging almost half of the village by 1984. In Kolma, Munshiganj, the villagers said that erosion is a regular feature, leading to permanent migration of many families to cities. They noted that while in 1987–1988, a few new *chars* (riverine islands) emerged, from 2004, erosion worsened, becoming devastating during 2006–2007. In the nearby Charipara, part of the village submerged during the mid-2000s, displacing 40 households to a neighbouring village, leaving only 100 families in the original place, as the villagers narrated.

A 60-year-old woman in Munshiganj narrated how riverbank erosion forced her to move to another village: "I am a direct victim of Padma riverbank erosion. I had a homestead of 24 decimal (0.24 acre) land and also three *bigha* (1.2 acres) cultivable land in Kolma union, which were inundated by the gusty river Padma in 1997. It took just two days for the river to swallow my homestead and a month for my cultivable land. Immediately after erosion, we moved to a place two kilometers away from what used be our home. Along with other, two (other) households we rented a 10 decimal (0.1 acre) piece of land; and together we pay the owner 9000 taka (GBP 74) every year. We built houses on this land with our money. Since then I have been living on this piece of rented land."

While riverbank erosion leads to permanent displacement, cyclones displace people in their immediate aftermath and then lead to seasonal migration as farms remain damaged and fallow after cyclonic storm surges. At Gabura in Satkhira district, the villagers said that Cyclone Aila of 2009 was the most devastating event in recent times. It flooded the whole village that is surrounded by rivers. In the *char* areas of Satkhira, the villagers said that Cyclone Aila left fields waterlogged for many months, leaving the soil and aquifers salinised, making fresh water scarce and farming impossible. A seasonal migrant in his late 40s from Satkhira district responded: "I am still struggling to find (a) livelihood. If (only) I work I can eat, but this is pitiful at this age. Riverbank erosion is taking away my land. Cyclone Aila destroyed my trees and resources. Now I do not have any resources." The perception of the villagers in Satkhira is that the frequency and/or intensity of cyclones has increased in the last few decades and their impacts are likely to worsen.

In all three districts studied, people have reported that they felt the temperature rising, especially during summer months. "This year, the heat is extremely intolerable," said a villager in Nawabganj. Many people also reported serious water stress. In Nawabganj, the villagers said that drought has become more severe over the past decade with, for example, a number of tube wells drying up. A farmer in his late 20s said: "The water level is going down day by day. Now water can be found 160 feet down. Ten years ago the layer of water could be found 100–120 feet down." In the *barine* areas, drought is a perennial problem. In Nawabganj, the villagers called water stress a 'crisis,' an expression that finds mention in policy documents (WARPO 2006).

The narratives show that people in different regions of Bangladesh perceive a wide range of climatic and environmental stresses and shocks, and they are concerned about the risks they pose. They are concerned about rising temperatures – though the rise has been gradual and small – uncertain rain and dipping groundwater levels. Salinity is becoming a threat that affects farming and drinking water availability in coastal areas. In some cases, such as riverine islands exposed to cyclones and erosion, people feel that such shifts are becoming more serious. There have been cases of reduced crop yield and crop losses. In short, people's narratives show a wide range of climatic and environmental stresses and shocks that affect their security, well-being and livelihoods. Changes and uncertainties pose further risks to their day-to-day lives and income prospects.

The following section examines to what extent these experiences and perceptions of risk influence the migration decisions of people.

4.6 Behavioural factors

4.6.1 Understanding migration decisions

In this part, I am trying to explore to what extent people acknowledge the role of climatic and environmental hazard experiences and risk perceptions in their migration decisions. The research focused on migration as a response of individuals, families and communities to a range of stimuli that affect their livelihoods, income, safety and well-being. The question is whether environmental and climatic factors make people move out of their place – at least temporarily. Is a flood or lack of rain for two seasons a good enough reason for someone to consider a change of residence? Do people abandon their place if they perceive it to be vulnerable to cyclones? To what extent are people concerned about the future risks posed by these hazards? These are some of the questions that we asked.

The answers that people gave acknowledged the presence of a range of hazards and their impacts on livelihoods. However, the general narrative was that yes, we have these problems, but we migrate to earn more money. Except in the case of people losing their houses or homesteads in riverbank erosion, the explicit connection between a climatic or environmental event and forced migration was rare. People did change their migration patterns though, especially after cyclones and the storm surges they bring along, rendering their fields saline and barren. Even then the reason generally given is livelihood, not the cyclone per se. So what drives this decision-making process? What are its stages?

To consider various stages and influences of the decision-making process, I analysed them using the model described above, taking into account a set of socio-economic and cultural factors along with climatic and environmental factors. The model disaggregates the behavioural components associated with this decision-making process into migration attitudes, migration control beliefs, migration personal norms and social norm beliefs.

4.6.2 Migration attitudes: migrating for better livelihoods and income

Migration attitudes denote how people assess various migration options before them and their efficacy – as a livelihood choice, a means to supplement income and at times leading to a safer habitat. The interview questions probed how the migrants and non-migrants decided to leave or stay at their village. The questions tried to elicit the respondents' experience of hazards, perceptions of risk, as well as perceived benefits brought about by migration, though not necessarily linking climate and environment with migration. Such an approach was necessary to prevent priming people on climatic and environmental influences, that means subconsciously influencing their answers.

Priming means preparing someone for a situation, typically by supplying them with associated information. In psychology, it means giving particular representations or associations in memory just before carrying out an action or task. Priming involves an implicit memory effect in which exposure to a perceptual pattern influences the response to another stimulus (Meyer and Schvaneveldt 1971). Experiments have shown that writing down the attributes of typical professors enhanced, and those of soccer hooligans reduced, respondents' performance on a general knowledge test (Dijksterhuis and Van Knippenberg 2008).

Bangladesh villagers make their migration decisions in their specific socio-economic contexts. It is often a means to improve income and offset losses suffered due to climatic and environmental stresses and shocks, overall contributing to resilience of rural communities. As for background, Bangladesh is a predominantly rural country with 54 per cent of rural work force engaged in farming and the rest in rural non-farm sector that is related to farming activities; so the transition from poverty in the country still has been dominated largely by higher income within the farming sector (World Bank 2012). Agriculture (including crops, livestock, fisheries and forestry) accounts for 21 per cent of the national GDP (World Bank 2012). Since the 1980s, however, structural changes in the economy have fueled urban growth, and the contribution of farming to the GDP fell from 30 per cent in 1990 to about 20 per cent in 2010; and the contribution of the urban sector to GDP increased from 37 per cent in 1990 to an estimated 60 per cent in 2010 (Muzzini and Aparicio 2013). For an average village household, that has meant the share of income from farming dropping from 59 to 44 per cent between 1987–1988 and 1999–2000, with services and remittances making up 35–49 per cent of income (Afsar 2003). Meanwhile, a labour force from the countryside drives city-based growth (Toufique and Turton 2002).

The government acknowledges a sharp increase in migration to cities amid rapid urbanisation (Planning Commission 2010). Even with a range of rural development measures that have made agriculture more productive and rural income generation activities diverse (Planning Commission 2011, 2012), migration continues in line with an international trend of diversification of rural livelihoods, including by increased mobility (Tacoli 2011). People living in climate-vulnerable regions, especially, try out secondary livelihoods that are not dependent on natural resources (Ahmad 2012). Faced with price hikes and wage drops during pre-harvest intervals, farm labourers often do other work or migrate to cities (Chowdhury et al. 2009), where jobs are easier to find than in villages (Afsar 2003).

In the cities, migrants often join factory jobs, engage in casual labour, open small shops or pull rickshaws. Women often migrate to work in garment factories. Such work in cities often gives adolescent girls a transition period from childhood, instead of early motherhood, as it might happen in a traditional village setting (Amin et al. 1998). In this context, Kuhn (2003) has identified two types of migration in Bangladesh. Poor village households send a family member to a city; or households that have lost their village-based livelihoods often move out to a new place, migration becoming an adaptive option in both cases. The argument here is that migration is a planned move, except in the immediate aftermath of extreme climatic or environmental events, when it could be a coping strategy. If it is a planned move for increased income in areas affected by climatic and environmental shocks and stress, it can be considered an act of adaptation.

In all three districts, the respondents said that the past four decades have seen a shift away from farming, traditionally the livelihood of a majority of the villagers. Their ancestors were farmers, who sometimes fished, gathered minor forest produce – such as bamboo, cane, palm leaves for thatching roofs, honey and bee-wax – from the Sundarbans mangrove areas of Satkhira and elsewhere. The new generation, however, has adopted a basket of livelihoods ranging from shrimp farming and vegetable selling to casual labour or small trade in a town or a city, just like Kazim from Gabura does. His father is still a farmer. Many villagers would migrate during the lean season or commute daily to a town. They would take up multiple roles as a farmer, rickshaw-puller, seasonal migrant, daily worker in town, toy seller and so on, depending on the season, the need for money and job availability.

Elsewhere, environmental changes coupled with economic opportunities have led to a shift away from agriculture, as the focus groups have revealed. In Munshiganj district, affected by frequent floods and riverbank erosion, such a shift can be clearly seen. At Charipara, for instance, the share of farmers in the village population has reduced from 60 to 30 per cent and fisherfolk from 25 to 15 over the past 30 years, as the villagers noted. In the nearby Mandra village, the number of farmers has "reduced drastically," as the local villagers noted. Elsewhere in the district, at Dakerkhati, 75 per cent of the people were reported to pursue farming, unlike 30 years ago when "everybody was a farmer." The devastation caused by the 1988 flood and continuing riverbank erosion that inundates fields further accelerate such a shift from farming. Erosion leads to displacement too. Charipara, a village of about 2000 people, has seen migration of 30 families because of erosion during 2007–2008, as the focus group discussion has revealed.

Migration patterns, however, differ across the districts. In the *char* areas of Nawabganj, migration has been a way of life, as the villagers narrated. In Chorpka, a village on the river Padma inhabited since 1980, for instance, annual floods often last three to four months and erosion affects farming, so job opportunities are limited. At least 10 per cent of the households depend on internal migration, as the villagers said. A 35-year-old woman here told her story: "In 1998 we were displaced due to flood and riverbank erosion. At first, we migrated from

Radhakantapur to Sohimullah village. Again we faced the same disasters in 2000, 2004 and 2008 there, so we came to Chorpka village." Her husband tried to start a ferry service in a village called Dinajpur, failed, moved to Chittagong and then moved to Mymensingh. "There is no other way to fight hunger," she said. "Besides, we need good savings for our children's study. . . . Marriage of three daughters is also a tension for both my husband and me." As per the prevailing local tradition, often the bride's family is expected to give a large amount – as cash or kind – as dowry to the bridegroom or his family.

In Chorpka and Durlogpur, villagers said they used to cross the Indian border five km away to work in rice fields or to trade in goods and cattle. They said that a border fence and tighter patrolling since 1995 have restricted this movement and that there has been an increase in migration to other districts within Bangladesh hence. In the *barine* highlands of Nawabganj, drought drives migration: "Economic hardship is the main reason behind my decision of migration. . . . In the village, in a year jobs are available only for 6 months in the agricultural land due to drought," a migrant farmer in his early 20s said.

However, most people do not acknowledge such a direct link between hazard risks and a shift from farming livelihoods. At Dakerhati in Munshiganj, a village that suffered floods in 1998, 2004 and 2007, a 30-year-old carpenter said: "I was unemployed before migration. My migration was driven by my desire to lead a better life, not by any natural calamity." In this village, over the past 10–15 years, internal and international migration has been a way of life. In the focus group here, however, people said they did not see migration as a solution to the problem of environmental hazards; except for landless labourers, who move to cities and towns during the lean season. Migration, instead, is seen as a way to improve the household income.

The narratives, however, do acknowledge the presence of climatic stresses and shocks, and the changing nature of weather patterns and their indirect influence on migration. For instance, a 42-year-old farmer from Nawabganj said that that rainfall from mid-July to mid-September has decreased in the last 10 years. "Due to lack of rain, the land cannot be irrigated. (We are) totally dependent on deep tube wells; even that cannot work during extreme drought season (summer)," he said. At least for six months a year he has to migrate to work as a rickshaw-puller. His reasoning is that it is economic needs that drive migration, though climatic factors influence migration decisions: "Economic hardship is the main reason behind my decision of migration . . . I migrate seasonally as jobs are not available here round the year. In the village, in a year, jobs are available only for six months in the agricultural land due to drought that puts my household under serious economic pressure. During the rainy season and winter, some work can be found in the village. This includes sowing seeds, and cutting paddy as day labourer." An inference can be made that even when migration is indirectly linked to the way people experience and perceive climatic stresses and shocks, they often do not make that linkage.

Such narratives show the difficulties many people face in earning a livelihood in marginal, hazard prone areas. Disasters dramatically disturb the precarious balance that they would have struggled to achieve, as a Satkhira farmer in his mid-40s narrated: "On 25 May 2009, at 3pm, the 7-foot-high tidal wave of Aila suddenly destroyed everything. My homestead, livestock, and all other goods floated away in the water. I took shelter in Khailasbunia School. After three or four months, I migrated to Sreengar *thana* (sub-district) of Munshiganj district with my two elder sons in search of livelihood. Now I work as a wage labourer, and seasonally migrate for work." In such scenarios, migration becomes a coping strategy.

The respondents of Satkhira acknowledged that the cyclone caused abrupt, unplanned movement. People had to leave their flooded low-lying villages, and move to other places on a higher plane, sometimes to clusters of temporary shacks built on embankments. Three years hence, at the time of the interviews, Aila's legacy lingered on, as the villagers could not grow paddy in their farms due to salinity left by the storm surge. It was a sudden shift in an already hostile environment. "Dramatic changes occurred after AILA of 2009," recalled a 45-year-old farmer and seasonal migrant from Sathkhira. "The village farms became waterlogged for two years. That made soil and water saline. All trees and all grass died. No vegetable or agricultural cash crop is grown here." The result has been devastating: "Food is scarce. We need to buy everything due to salinity of the soil . . . poverty, monetary crisis, scarcity of fuel to cook . . ."

Three years after the cyclone, men spent winter months, when there is no farming in the area, migrating to towns and cities. Earlier, only 10 per cent of the people migrated, but after the cyclone, 50–60 per cent of people moved for temporary or seasonal work in other districts, such as Gopalganj, Jessore, Khulna, Magura, Bagerhat, Madaripur, Munshiganj and Dhaka, the villagers said. A government ban on local shrimp farms – because it damaged the environment – and irregular rainfall have also contributed to an increase in migration, the villagers said. Some shrimp farmers had let in saline water by breaching the embankments, leading to an even worse impact due to the 2009 storm surge, the villagers added. This violation lent credence to the environmentalist's call for a ban on aquaculture in the area. Some of the responses shared visions of a bleak future for the place: "Due to regular floods people almost cannot do anything during that time. It offers no opportunities to do any business."

The distance and duration of migration differ across the study areas. Migration could be internal to Bangladesh, as in most of the villages studied, or international, as it is noticed sometimes in Munshiganj. In Charipara village of Munshiganj, 10 per cent of people work abroad, mainly in the Gulf countries, Singapore and Malaysia according to the villagers. In Mandra, about 40 per cent of the households have one or more migrants working abroad. At Bhagyakul, in the same area, 50–60 per cent of the households have an international migrant, the local villagers said. In Mandra, the villagers said many of them commuted to Dhaka, the capital city that can be reached in 20 minutes by bus. Some people migrate to work in garment factories in Gazipur and Narayanganj districts, networking through those already working there. From Bhagyakul, some men migrate to Chittagong to sell utensils and silver and plastic ware. While internal migration from Munshiganj is not as pronounced as in the other regions, the district also serves as a destination point for migrants from the coastal belt of southwestern Bangladesh as well as the drought-prone northern districts such as Kurigram and Rangpur (Ahmad et al. 2012).

The narratives show that people view migration as an activity that contributes to better livelihoods in a context of exposure to climatic stresses and shocks. The results in general suggest that migration is seen primarily as an economic activity, a livelihood choice. In many cases, climatic and environmental stresses and shocks work in the background making livelihoods difficult and destroying habitats and means of livelihoods. However, except in the case of disaster-induced displacement, people tend not to associate environmental and climatic stresses and shocks directly with migration.

4.6.3 Migration control beliefs: expectations and perceived resources

The above section shows that migration decisions tend to be adaptive and deliberate; however, the migrants do not always believe that they are in full control of their situation. Simply put, migration is often not seen as a certain choice for a better life in the face of uncertainties and adversities involved in local climate and environment. For the villagers, migration decision-making involves a set of uncertainties and limited information about a wide range of factors – scope for farming, changing global markets and financial strain, besides environmental and climatic stresses and shocks. In this context, migration control beliefs comprise generalised expectations about the extent to which people think they can control events that affect them and the perceived resources available for migration.

Subjective or perceived adaptive capacity, or what an individual or a community thinks it can do, given the availability and access to resources, is as important as objective adaptive capacity, or what can actually be done (Grothmann and Patt 2005). Specifically, studies in the delta areas of Bangladesh have shown that a belief that disaster occurrence is in the hands of God does not prevent people from preparatory action (Alam and Collins 2010). Psychologists would explain it in terms of what they call the locus of control. It denotes expectations about the extent to which people think they can control events that affect them (Rotter 1966; Price and Leviston 2016). An internal locus of control denotes people's expectation that they could be in control of their future. An external one would show their belief in some other force controlling their future.

A confirmation of an external locus of control among the respondents was the predominant agreement to the statement in the interview questionnaire: "Many times I feel that I have little influence over the things that happen to me." A usually shared comment was variations of the notion that "God determines everything." The question is whether people feel in control of their destiny under such circumstances marked by uncertainty and do not feel as if they were mere pawns in the hands of fate. Interviews revealed a rather nuanced pattern of control beliefs. Many of those interviewed said that the success of a household lies mostly determined by factors outside of their control, suggesting an external locus of control.

Most of the interviewees also agreed or strongly agreed to a related statement, "No matter what things I try to make a living in my village, the drought/flooding etc. prevents them from working." They tend to agree more or less with all the locus of control statements except the one that said despite short-term difficulties due to weather and commodity prices, an individual can stay ahead in the game. Only this result suggests an internal locus of control. However, in the context of adaptive action taken by the people – namely the wide basket of livelihood activities, different forms of migration and so on – it seems this sense of helplessness does not prevent people from taking decisive action.

Migration control beliefs determine the way people make their decisions to stay put in their place or move out in the face of economic pressures in a background of multiple climatic and environmental hazards. A farmer in his mid-40s from Satkhira collecting forest produce, selling vegetables and groceries and migrating, also expressed similar sentiments. Three months after Cyclone Aila, he and his two sons migrated to Munshiganj, looking for wage labour. His father had sold his 33-decimal (0.3 acre) land to meet family expenses. "I tried many ways to become successful in different livelihood activities, but failed. Natural calamities are also a big reason for this failure along with human-made policies," he said. In the drought belt of Nawabganj, a migrant rickshaw-puller in his early 40s said: "Crops often failed in the drought season. My father was a sharecropper. However, due to lack of irrigation facilities he could not grow crops round the year. Over the years, he failed to pay his dues to the landowner, was burdened with debt, and then he started working as a day labourer. I also do the same."

Even while narrating these stories of helplessness in the face of an uncertain climate, meagre resources, inadequate infrastructure and the lack of any social safety net, the respondents revealed the power of human agency in taking effective adaptive action such as migration. Migration is a choice they make in the face of adversities or opportunities. For a 30-year-old respondent in Munshiganj, leaving the farms in his flood-prone village to become a carpenter was clearly a choice for better earning. "My migration was driven by my desire to lead a better life, not by any natural calamities," he said. In the way people frame the narrative of climate, environment and migration, the focus group discussions and interviews reveal a certain 'can-do' spirit, despite seemingly insurmountable obstacles. To put it concisely, the respondents see migration as an act of agency, a positive, planned move they take for a better livelihood. In this framing, migration is a not an act of helplessness by people faced with climatic and environmental stresses and shocks.

4.6.4 Migration personal norms: self-concept as opinion leader, perceived level of risk-taking

Migration personal norms denote to what extent people believe that they are change-makers and opinion leaders, who take risks and innovative approaches. This section probes the migration decisions even further to see to what extent self-concept as an opinion leader and a risk taker influences migration decisions of the respondents. This is a logical follow up of the last sub-section that conceptualizes migration as an act of agency.

The interviews showed that migration is a display of agency by villagers who wanted to earn more or offset losses suffered because of environmental changes. All the interviews suggest that it is the individual migrant who makes the migration decision. However, there were consultations with family members in the decision-making process, and support for migration came from extended family and friends, as explained in detail in the following section. In the case of a 30-year-old man, his mother played a role in decision-making, and for a 25-year-old man, his father and brother contributed to the process. Two women interviewees said their husbands were the migrants and the two men took the decision to move.

Only five out of the 20 respondents considered themselves to be among the first in their area who have changed livelihood options, showing a level of pioneering spirit in the face of adversities or perceived inadequacies. A 32-year-old migrant in Nawabganj said that his father and ancestors were farmers, but after his homestead and farms were eroded, he moved to another village 23 km away. "As I do not have any land and farm work cannot be found all the time, I started working outside as a farm labourer, and later as a hawker in other districts of Bangladesh."

However, most of the interviewees did not consider themselves to be among the first to embrace change. A majority (11) of the respondents still said they were trying new livelihoods. Migration is not seen as a pioneering or risk-taking venture, but a business-as-usual activity despite all the uncertainties involved in it. The question whether they consider themselves as more risk-taking than others elicited no or negative response from all participants, except one. A 41-year-old landless labourer who used to cut and sell mangrove forest trees said: "Yes, it may bring fortune." He said he continued his father's trade and now has become a labour contractor.

At the same time, the respondents appreciated the risks and hardship involved in migration. Most of them said they could not take their family to their work destinations due to a variety of reasons including the temporary/seasonal nature of migrations, social commitments back home and possible exposure to risks. As a contractor who supplies labourers from his village to a brickfield in Satkhira district summed up the reasons for not taking his wife and two children with him: "It will be risky if we face any bad situation; social and religious practices; affection to village."

The narratives of migration also show that people test and tweak their methods based on their own and their peers' experiences. Interviews suggest that it is usually informal networks that recruit and sustain migrants. As one of the respondents in Nawabganj narrated: "Before migration, there was no work in the village. I used to roam around here and there in search of livelihood. . . . First, I went to Katapukar, another village. . . . There I met a day labourer named Sadikul, who first told me about rickshaw pulling in Rajshahi." He also used to work as a seasonal rickshaw-puller in Rajshahi. "Now I work six months in the village in the agricultural land and the rest of the time in Rajshahi." Most of the migrants (13) were "sometimes" consulted by others on issues regarding migration. One each was consulted "frequently" and "all the time." Five of them had "moderate" and three had "significant" influence over the livelihood practices, including the migration of others.

Together, these responses suggest that migration decisions are often made independently and the migrants are open to new livelihood options and moving to new places. Though migration is considered a "new" occupation as opposed to what the migrants' fathers and ancestors did, it is not considered a particularly risky or unique venture. Under the changed economic and environmental circumstances, there is a business-as-usual sense to migration despite the uncertainties involved in it. Migration appears to be a reliable option despite the uncertainties and a lack of any formal support by the government.

4.6.5 Social norm beliefs: sources of advice, traditions and cultural factors

Social norm beliefs explain the perceived trust in and influences of sources of advice, traditions and cultural factors. A social norm can be defined as "a rule governing an individual's behavior that third parties other than state agents diffusely enforce by means of social sanction" (Ellickson 2001:3). Social norms are analysed here to further understand influences involved in migration decision-making. If it is not primarily climatic and environmental factors or a spirit of risk-taking that drives migration, there are other factors at work here. One such factor cited in the literature is social traditions and customs that legitimise migrations and support it.

It has been argued that the behaviour of one's peers, colleagues and family members influences one's identity and behaviour; people also learn to value something through their own experiences (Kinzig et al. 2013). Hunter and David (2011) argue that culture-specific norms shape the ways in which households diversify livelihoods, including by migrating in a changing climate. These include gender norms. Women often become disproportionately vulnerable to natural hazards. Gender inequality and family responsibilities limit their mobility and survival options (Ahmad 2012).

In making livelihood choices, people put their trust in fellow households, as the qualitative study has revealed. NGOs and the national government also enjoy their trust, but on a lesser level. Most of the respondents said that their decision to change their livelihood was influenced – "a little" (four respondents), "moderately" (six) or "significantly" (two) – by the behaviour of their neighbouring households, friends or family. However, seven respondents said that there was no such influence. The migrants trust their social networks to inform them about opportunities and places to migrate. Usually when people go to work outside the village, family members, relatives or their friends share notes with them, and they take this as the primary source of information and advice. Interviews suggest that people trust these informal sources much more than government agencies and institutions.

The story that a widow in her mid-50s from Munshiganj narrated weaves in the bit that social networks played in her son's migration abroad: "My son was unemployed and he was not interested in farming. Some of these friends from this village and outside have migrated abroad. He tried to convince me in different ways. However, due to financial constraints, I could not agree with him. Later he came up one day with a sub-agent who facilitated migration of many people from this village. The sub-agent told me about a job abroad, salary and other benefits. Knowing everything, I went to my brother, who is financially in a better position, and asked for loan from him. He agreed to give me half of the migration cost. I also had a contact with an NGO in other village, who gave me one-third, and the rest of the money my son secured from one of these friends."

People depend on social networks for advice on livelihood activities as well. Regarding farming practices, 19 out of the 20 respondents declared "complete trust" and the remaining person "trust" in advice from their fellow households. Only half the respondents placed trust in information given by national and local governments on this matter. At the same time, 13 respondents said they were influenced by the behaviour of neighbouring households, friends and family in their decision to change their livelihoods; eight respondents said such influence only had a moderate effect, and one said it had a significant influence. Ethnic ties also often play a key role in facilitating migration. In Gabura, a woman in her mid-20s belonging to a minority ethnic group said that her husband was paid less than what his colleagues got – now he migrates to town with fellow villagers led by a labour contractor, all of them belonging to the same ethnic group.

Another culturally significant feature noticed in migration patterns is the prevalence of male migration. Often women and children cannot accompany male members of the family to migrant destinations. While it suggests that migration could often mean roughing it out in hostile environments, it also means that women prefer not to move to a new place without adequate facilities. At the same time, in some cases women prefer to stay back in risky home environments, especially after disasters such as cyclones (Mallick and Vogt 2012) to head households and take care of local livelihoods under trying circumstances. As the Gabura respondent belonging to the minority ethnic group narrated: "My husband is now working as a seasonal migrant worker. . . . His seasonal migration has turned into (a more frequent, but) temporary form immediately after Aila, with the closure of shrimp farms for two years and the restriction of government on pursuing livelihood in Sundarbans. He works in Jessore, Khulna, Gopalganj and Satkhira in the brick and paddy field. Like many other women (here) I catch small shrimp and crab from the river and sell them to small (retail) buyers."

These narratives of migration suggest that people follow their family members, peers, friends and community members while choosing their migration paths. Focus groups and semi-structured interviews showed that migrants placed a high degree of trust in information from their social networks while deciding where and when to go and what to do for a living (Massey et al. 1993). Usually, people who also work outside the village, family members, relatives or their friends provide information about migration. Mostly, resources for migration come from family members. In terms of social norm beliefs, the migrants trust members of their fellow households the most. Overall, socio-cultural norms and beliefs play a key role in making migration decisions. This finding has policy implications, as it is word-of-mouth sharing of experiences among peers and relatives that often influences migration decisions, not formal institutions.

4.7 Discussion

Research into the nexus between climate change and migration has often focussed solely on how people move in response to the impacts of variability and change in climate. This notion often ignores the nature of migration as a tried and tested livelihood choice amid a variety of socio-economic and environmental opportunities and limitations. This chapter closely looks at the behavioural aspects of migration decision-making in Bangladesh in the context of changes in its economy, and, increasingly, exposure to the impacts of climate variability and change.

The chapter traces the way migration decisions are made in the context of environmental change, including the impacts of climate variability and possibly change. Faced with dramatic changes, people are diversifying their livelihoods from the farming and fishing that their ancestors practiced. They migrate to become shrimp cultivators, vegetable vendors, rickshaw-pullers, street-sellers, casual labourers, contractors and factory workers. Depending on the availability of jobs, migrants often take up different roles, earning from a basket of livelihoods in any single year. Against a background of economic growth and reduction in farm livelihoods, villagers are confident about making use of the emerging opportunities in cities. They are positive about the efficacy of migrant labour as a way out of the limited job opportunities and sometimes losses suffered because of climate- and environment-related stresses and shocks back home.

People see migration – in different patterns across time and space – usually as a strategy to diversify livelihoods. Sometimes migration is a coping strategy after a climatic or environmental shock. Households often diversify livelihoods by sending one or some of the household members away to work – for different durations – and thus reduce their vulnerability to shocks and stresses, including climatic ones. At the same time, the prevalent narrative about migration is all about improving income, and not necessarily escaping from a hostile environment, even when environmental stresses and shocks make livelihoods increasingly insecure and unsafe. In short, in a range of time–space combinations, migration contributes to such efforts, though migrants themselves do not call it adaptation.

In this context, migration decisions are often taken firmly and deliberately. Even when there are climate- and environment-related threats, the decision-making process involves weighing the pros and cons of migration against other options such as diversifying livelihood activities at the home base. Though villagers tend to believe that the success of their household is mostly determined by factors outside their control, their creative and bold adaptive actions suggest that they have a sense of control over their destinies. Their belief that disaster occurrence is in the hands of God, however, does not prevent them from taking preparatory and remedial action. Still it may be noted that migration is not always possible or feasible due to a variety of reasons, including lack of financial resources, family commitments and inadequate facilities and networks at migrant destinations.

The research shows that villagers in areas particularly affected by increasing climatic stresses and shocks are diversifying their traditional livelihood strategies by migrating. Environmental factors, including climatic stresses and shocks, often make such shifts even more necessary. Although the migrants' primary motivation is better income, in effect, migration becomes an effective form of adaptation.

4.8 Conclusion

Based on a qualitative study in three geographically distinct places of Bangladesh, this chapter proposes that migration is a socially acceptable behaviour that occurs in the context of perceived environmental change and climate variability. Migration decisions are mediated by a set of behavioural factors that assess the efficacy of different responses to opportunities and challenges, their socio-cultural acceptance and the ability to respond successfully. This is because proactive action against natural hazards requires more than just risk awareness; it also involves helping people cross barriers to adaptive behaviour and promoting social settings and environments that allow responsible action (Grothmann and Reusswig 2006).

A nuanced understanding of migration decision-making is particularly important considering the urgency for climate change adaptive action in Bangladesh. Firstly, climate extremes and even a series of non-extreme events are occurring against a background of social vulnerabilities and exposure to risks (IPCC 2012). While this trend continues, it is important to understand, model, forecast and disseminate information on how the climate is varying and changing in the long term and how people are responding to such changes – by moving, staying or getting trapped. Secondly, the economic impacts of climate-related disasters on livelihoods continue to be huge. Entire stretches of land are still being eroded, salinised, flooded or kept fallow due to water shortage. People often have to move out of their place of origin for their safety and due to limited livelihood opportunities – but they also run the risk of a new set of hazards and uncertainties in their destinations. An understanding of the migration dynamics and patterns could help in planning for future development and resilience of migrants as well as their home and host communities. Thirdly, while migration works as an effective adaptation strategy to address both current and future environmental stresses and shocks, it is seen as a business-as-usual economic activity by most of the migrants. People in the study areas see migration as a way out of economic difficulties and expect environmental conditions to worsen in the coming decades, which means possibly more migration. If migration is an effective adaptation strategy, it is important to mainstream it into development, climate change and environment policies.

The next chapter tests how climate- and environment-related hazards relate statistically with long-term migratory movements. Such a nuanced understanding of how environmental concerns influence internal and international migration has policy and development implications (Hugo 1996; World Bank 2010).

5 Climate, environment or poverty?

Statistical analysis of factors that drive migration

Following up on the qualitative evidence, this chapter numerically tests how economic as well as climate- and environment-related drivers work together in influencing migratory movements. Based on logistic regression models and event history analysis of data from 1386 respondents from the study areas, it tests the factors that influence their first migration outside the district and the first house shifting, that is mostly within the district. The evidence suggests it is predominantly income needs that drive migration, though those without any assets are often unable to migrate outside the district. It is respondents in their 20s who migrate more than others, aided by their education and social networks. Experiences of drought and cyclone tend to positively influence migration outside the district, but riverbank and coastal erosion negatively. Rainfall uncertainties tend to influence migration too, both positively and negatively. The findings are explained and placed within the context of similar recent studies on climate- and environment-related migration in Bangladesh.

The previous chapter has shown that in the three study districts of Bangladesh – Nawabganj, Munshiganj and Satkhira – villagers experience climate- and environment-related stresses and shocks, and they are concerned about the risks they pose to their livelihoods and security. However, when people migrate, they rarely attribute their decision to move out of a place to experiences of hazards or risk perceptions. Instead, migration is seen more as a business-as-usual activity aimed at generating better income, offsetting losses or rebuilding after disasters. Climate- and environment-related factors work in the background at best, without meriting an explicit acknowledgment. This chapter further explores such ambiguities involved in people's attribution of climate- and environment-related factors to their migration decisions. Based on field survey data, it develops a statistical model to explore how experiences of climate- and environment-related stresses and shocks might influence migration – even if the respondents do not explicitly mention such a connection.

This chapter follows up on the evidence presented in the previous chapter, using survey data from Nawabganj, Munshiganj and Satkhira districts. While economic as well as climate- and environment-related drivers work together in influencing migratory movements, it is not always clear from the qualitative enquiry how they interact with one another. A statistical model can help trace these different influences and their overlapping effect on migration. Using an event history analysis approach, this chapter builds a main logit model and a set of supplementary models to understand the dynamics of climate- and environment-related migration. The assumption here is that the first migration influences subsequent movements. Therefore, the main models look at the respondent's first movement outside the district – irrespective of the destination. It tests the sensitivity of these movements to their environmental and climatic experiences and compares them with socio-economic factors that also influence migration.

The first part of this chapter spells the methods used for quantitative analysis in direct follow up of chapter 4. The second part gives a snapshot of the dataset and its characteristics. The third part looks at the models in detail and the fourth part summarises and discusses the results. With the aid of tables and graphics, this chapter analyses how first movements – within and outside the district of origin – are influenced by various climatic and environmental experiences; and how migration trends differ when different socio-economic parameters are taken into account.

5.1 Methods explained

5.1.1 Introducing the regression model

To understand the factors behind first migrations in Bangladesh, this thesis adopts a discrete event history model, analysing it using the statistical analysis software, Stata. Event history is a longitudinal record showing when a sample population experienced one or more events. Event history analysis is often used to study the duration until the occurrence of the event of interest. The duration is measured from the time at which an individual becomes exposed to the 'risk' of experiencing the event. A set of explanatory variables are considered as potentially influencing the risk exposure. Two of its features, namely time varying variables as well as censoring, that is removing a subject after the event, make it difficult to be analysed using standard statistical procedures (Allison 1982; Steele 2004). Event history analysis methods can help identify causes of events.

In this research, the first migrations and the influences of socio-economic variables and the influence of a set of climate- and environment-related events are considered. It involves a person-year data structure, and data censoring occurs due to first migration of the subjects. Event history analysis is well suited to analyse such data with such a design (Allison 1982). In line with the literature, binary and multinomial logistic regression methods are used to build discrete-time event history models (Henry et al. 2004). The logit model comprising individuals at risk of migration can be expressed as:

$$\log\left(\frac{p(y_t = 1 \mid x_{it})}{1 - p(y_{it} = 1 \mid x_{it})}\right) = \alpha_t + xt\,\beta$$

Individuals at risk (of migration) have been included in a logit model in which $P(Y_{it} = 1|X_{it})$ is the probability that individual i experienced a first internal migration conditioned on a set of variables that took effect in a time span of t. The set of variables includes household and individual characteristics, experiences of climate- and environment-related hazards, concerns about these hazards and rainfall data. This set of variables includes time invariant and time varying variables. The baseline hazard function is α_t that is specified as the log of time spent in the risk set (simply put, time prior to the migration event occurring). The odds ratio can be written as the ratio of the probabilities that the migration event occurred to the probability that the migration event did not occur:

$$\frac{p(y_t = 1 \mid x_{it})}{1 - p(y_{it} = 1 \mid x_{it})} = e^{-xt\beta}$$

In event history analysis, hazard denotes the probability that an event occurs within a very small interval of time given that an event has not already occurred. Here the interval is a year. The data in this chapter has been transformed from odds to log of odds (log transformation). This is a monotonic transformation; i.e., the greater the odds, the greater the log of odds and vice versa (Cooke et al. 2013). In log of odds form – as used in this chapter – if the coefficient is positive then the factor of interest raises the hazard of the first internal migration occurring. A negative coefficient indicates that the factor reduces the hazard of first migration.

5.1.2 Selecting the variables

The literature suggests that a set of socio-economic variables are expected to have an influence on the decision to migrate. Mainly they include gender, age, education and level of poverty, assets owned, social networks and family size. Women and men often have different migration behaviours mainly because of the different economic and social roles they play in the rural economy, and their job prospects in cities, as explained in chapter 2. The literature justifies disaggregation by gender to accommodate differences in migration patterns (for instance, Henry et al. 2004and Henry and Dos Santos 2012). Climatic and environmental stresses and shocks – especially in the context of climate change – can affect women's and men's assets and well-being differently with regard to farm production, food security, health, water and energy resources, climate-related disasters and migration (Goh 2012). Social and cultural norms that determine gender roles and women's lack of ownership and control over assets make them more vulnerable, as the literature shows (Goh 2012).

Besides, men tend to explain their migration in terms of farming or financial needs, while women respond in terms of family reasons. The literature notes that in Bangladesh often seasonal or even long-term labour migration is predominantly a male activity (Afsar 2003; Chowdhury et al. 2009; Mallick and Vogt 2012). Therefore, regressions have been carried out separately for men and women after building the common logit models that consider the research questions.

The main logit model, however, includes men and women, because migration comprises men and women and a true picture of its relationship with various socio-economic variables can be understood only in a common model. To a lesser extent, such a common approach makes sense because migration is often a family activity, the different members or the whole family participating in it, supporting it or contributing to the migration decision-making process, as the qualitative analysis has shown. Besides, the sample of inter-district migration is diluted with gender-based segregation possibly not capturing the whole picture of migration. To give an indication of different profiles of migrant women, hazard ratio and survival analysis graphs have been attached to supplement the main models. The separate, gender-based logit models are therefore given as supplementary data.

There are other socio-economic factors included in migration models in the literature. Age could be an important factor in determining migration. The literature suggests that migration decisions are influenced by the stage of life of the migrant when he or she moves. In Bangladesh, it is often younger people who migrate in search of work, as the qualitative research shows, and labour markets prefer younger recruits to older ones. The literature shows education as a key variable influencing migration in Bangladesh (Haque and Islam 2012). The rationale is that education opens up avenues, enhances skill sets and gives a broader worldview, thereby increasing the chances of getting the respondent a place in the labour market. As migration represents an effective risk mitigation strategy (Halliday 2008), large family size often places more demand for resources and, therefore, often migration. On the contrary, sometimes family size can also act against migration due to the costs of travel. Family size and number of children in the family were, however, omitted from the final models. They did not show significance across regressions, and they did not provide consistent model results.

As for networks, migration of a family member allows pooling and reduction of risks. Similarly, the decision to migrate is more likely when individuals have migrant networks present in the destination (de Haas 2010) and/or sufficient income to support their journey and meet the initial expenses at the destination. The qualitative analysis shows how migrants depend on their social networks for information, inspiration and financial support.

The literature shows that migration is often driven by the need for better income, to offset or minimise risks and to recover from losses (Stark and Levhari 1982; Stark 1984; Stark and Bloom 1985; Massey et al. 1993). Migrants look for rewards at their destinations (De Jong and Fawcett 1981), or an escape from disruptions or reduction in income (Stark and Levhari 1982). Migration still is often determined by socio-economic status (Wolpert 1965). Often a migrant's wealth, including money to undertake migration, influences his or her migration (Skeldon 1997). Among the socio-economic variables, one that is closely related to income is assets.

Along with income (expressed as poverty or lack of sufficient finances), the number of assets have also been included as key socio-economic variables in this analysis. Though people draw on natural, social, human, physical and financial assets (de Haas 2008) for migration, the asset variable in this analysis only represents the last two. The other aspects have been considered under social networks and environmental and climatic variables. Migration literature places assets as an influencing factor in the unique combination that determines migration decisions (Kniveton et al. 2011). However, it may be noted that poverty and assets can also have opposing impacts on migration, cancelling out the influence of each other. The literature shows that it is not always the poorest people who migrate (De Haas 2007). While assets aid migration, presence of assets can also work as a sign of affluence that makes migration unnecessary. Some people, under certain circumstances, may be too poor to migrate.

Climatic and environmental stresses and shocks appear to increase short-term rural-to-rural migration, but often do not affect, or even decrease long-distance moves. Henry et al. (2004), for instance, found that drought in Burkina Faso increases such short-distance moves while decreasing long-term, long-distance, mostly international moves. Often in Bangladesh, men often migrate in the event of a cyclone, leaving behind the family (Mallick and Vogt 2012). Such temporary migration of men might not count as shifting residence; however, cyclone-affected coastal regions show higher rates of migration compared with other parts of the country (Marshall and Rahman 2013).A large part of migration in Bangladesh also comprises people trying to escape seasonal deprivation (Chowdhury et al. 2009), especially from the northern drought belt, and to recover from the impacts of natural hazards (Hunter 2005; Penning-Rowsell et al. 2013). If these temporary moves do not involve a change of residence as such, they are not captured by this analysis, as explained in chapter 5 as part of the limitations of the methods.

However, climatic and environmental hazards – such as cyclones, floods and drought – do not necessarily make people move out or migrate in large numbers, except in the case of riverbank erosion or salinisation of farms (Gray and Mueller 2012; Penning Rowsell et al. 2013; Bohra-Mishra et al. 2014). While migration helps people cope with climatic and environmental stresses and shocks, literature also shows that disasters can sometimes reduce migration by cutting down resources needed to migrate or by increasing labour needs in the points of origin (Gray and Mueller 2012).

To address environmental factors as possible causal influences on migration, this analysis includes two distinct sets of variables: first, self-reported experiences of floods, cyclones, erosion and droughts in each place of residence; and second, measured rainfall data obtained from meteorological stations in each location. Observed rainfall data from meteorological stations located in the three study districts (since 1981 until 2012, the survey year) were collected to measure the rainfall variability alongside the self-reported data of other climate- and environment-related hazards. The literature suggests that there has been an increase in rainfall variability throughout the April–October season, and a shift in distribution, a reduction of overall rainfall and intense rainfall in October (Ahmad et al. 2012). Such variability disproportionately affects poor farmers with small land holdings and fishers by changing flood patterns (Ahmad et al. 2012). As a proxy for rainfall variability, anomalies in rainfall have been taken as the variable for regressions. Besides, observed data on floods and cyclones, including the extent of flood area, its percentage, people affected, as well as cyclone wind speed, intensity and storm surge height, casualties and people affected were collected and tested for their correlation with migration.

5.1.3 Data quality control

Individual correlation tests were done for a set of socio-economic and climate- and environment-related variables selected from the survey. The quantitative survey questionnaire attached to this document as annex 2 gives a full list of the variables used in the survey. Based on the preliminary correlation tests, a set of variables that are correlated with migration are taken to build an event history model. Variables that show no correlation in the initial models were dropped from the final model. Observed flood and cyclone data were dropped from the final model, as they appeared to be too coarse to identify district-level migration trends. Concerns of risk posed by the above hazard experiences – as expressed in the qualitative analysis – were also tested for correlations. However, when incorporated into the models they appeared to be too nuanced to elicit consistent results across the sample population. It may be noted that answers to questions about risk concerns gave an indication of how people respond to risks in the qualitative part of the study.

Finally, the following socio-economic variables were included in all regressions reported in final model control for the various household characteristics that affect first migration: age at migration, education of the respondent, poverty while at the place of origin, number of assets owned at the place of origin and family and friends outside the district of origin. The environmental variables considered were rainfall anomaly for each year of stay at the respondent's place of origin based on observed station data, and self-reported hazard experiences at the place of origin of the respondent, namely droughts, floods, riverbank erosion and cyclones.

An issue in estimating the regressions is unobserved heterogeneity referred to in the event history literature as frailty. Given that the data on individuals in the dataset are in panel form, there might be unobservable factors determining their propensity to migrate. One way to control for frailty is by assuming that the data is normally distributed – this can be done by estimating a random effects logit. The random effects logit assumes that the unobserved heterogeneity is normally distributed and all estimates are conditional on this distribution of unobserved heterogeneity. All tables in the paper are based on a pooled logit whereby no assumptions have been made about the distribution of any unobserved heterogeneity present in the data.

Logistic regression could involve errors, including measurement problems due to inaccuracy of the instruments used and the subjective nature of certain variables, such as self-reported information. Measurement error correction techniques have been suggested in the literature, but most of them require making certain assumptions on the involved variables, and usually it is very difficult to check whether these assumptions are satisfied, mainly because of the lack of information about the unobservable and mismeasured phenomenon. Therefore, to achieve better robustness in analysis, alternatives based on weaker assumptions on the variables may be preferable (Guolo 2008).

One option to enhance robustness is variance – covariance matrix estimation (VCE) corresponding to parameter estimates. The standard errors reported in the table of parameter estimates are the square root of the variances of the VCE. Standard error of estimate is a measure of how much each data point on an average differs from the predicted data point. It is like the standard deviation of all the error scores and informs how much imprecision there is in the estimates calculated (Salkind 2010). This robust estimator of variance can relax the assumption of independence of the observations. In Stata it is done by the VCE (cluster id) option, producing "correct" standard errors (in the measurement sense), even if the observations are correlated. It specifies that the standard errors allow for intra-group correlation, relaxing the usual requirement that the observations be independent. That is, the observations are independent across groups (clusters) but not necessarily within groups. A cluster variable command (clustvar) specifies to which group each observation belongs, for example, VCE (cluster id) as used in this chapter accounts for observations on individuals. Clustering produces a valid inference whether or not heteroscedasticity or autocorrelation are problems.

In all regressions the baseline hazard was significantly different from zero and positive, indicating positive duration dependence (the hazard increases with time). Besides, care has been taken to avoid multi-collinearity by introducing variables one by one in the model in a stepped manner, and isolating variables that might work together or cancel each other out. In the event of suspected multi-collinearity, alternative combinations were also included in the model to offset it.

Further, it may be noted that only about a fourth of the first movements of an individual were outside the district, as the summary statistics show. Therefore, to test what exactly drives inter-district movement, and how an instance of inter-district migration varies from a shifting of house within the district, a separate logit model has been built comprising the first instance of house move irrespective of the destination. This model gives an indication of how the same socio-economic and climatic and environmental factors that determine the first inter-district migration influence the first house moving as well. This may be seen as an additional tier of analysis that gives more depth to the understanding of factors behind climate- and environment-related migration.

5.2 Descriptive statistics

As the objective of this quantitative analysis is to explore the relationship between migration and a set of socio-economic as well as climate- and environment-related variables, the key dependent variable modelled is first migration. Migration has several definitions in the literature. For this analysis, 'migration' is taken to mean any move that an individual makes from his district of origin. The initial movement from the district of origin is labelled here as 'first migration.' Subsequent migrations to other places have been excluded from the model, as migration is in part path-dependent, such that first migration might partly explain subsequent movements (Balaz and Williams 2007). In the event history dataset used for logistic regressions, each year of the stay of a respondent in his first place of residence has been listed in a row. Each observation denotes a socio-economic condition or experience of one of the climate- and environment-related hazards, or rainfall condition based on actual station data.

Table 5.1 gives the basic details about the sample and its composition in terms of gender, location and district of origin.

Further, Table 5.2 gives a break up of respondents by their district of origin and whether they have moved houses and their first move involved migration outside

Table 5.1 Respondents who have migrated outside their district of origin

	Frequency	*Percentage*
Total sample size	1386	100
Non-migrants	633	45.67
Migrants	753	54.33
Women	464	33.48
Men	922	66.52
Migrant women	140	
Migrant men	613	
Sample after censoring outmigration by respondents under 15	1317	
Migrants aged 15 or above*	686	
Female migrants aged 15 or above	128	
Male migrants aged 15 or above	558	
Sample size by interview venues		
Dhaka	184	31.28
Munshiganj	272	19.62
Khulna	179	12.91
Satkhira	293	21.14
Nawabganj	458	33.04
Sample size by district of origin		
Munshiganj	454	32.76
Satkhira	472	34.05
Nawabganj	460	33.19

* Logit model for inter-district migration excludes migrants below the age of 15

Table 5.2 Respondents who have moved houses

District of origin	*Respondents*	*Non-movers*	*Movers*	*Move outside district*
Munshiganj	454	8	446	80
Satkhira	472	103	369	109
Nawabganj	460	73	387	120
Total	**1386**	**184**	**1202**	**309**

Table 5.3 Migrants and non-migrants by their district of origin

District of origin	*Non-migrants*	*Migrants*
Munshiganj	238	216
Satkhira	207	265
Nawabganj	188	272

the district. It can be seen that the vast majority of the respondents (1202 out of 1386) have moved houses at least once.

Table 5.4 shows the major first destinations of the respondents who have migrated out of their district. It may be noted that the destinations are big cities, topped by Dhaka, followed by Khulna, a coastal city.

Table 5.4 Major destinations of the migrants

Destination district	*Respondents*
Dhaka	391
Khulna	201
Chittagong	23
Rajshahi	21
Jessore	17
Nawabganj	11
Other districts	90
All districts	**753**

5.3 Understanding the variables

Among the socio-economic variables, age has been taken at the time of migration, as it is a key factor that influences migration decisions. For migration outside the district, all the respondents who were less than 15 at the time of the migration have been excluded. This is to ensure that the model represents an act of migration undertaken by the respondent and not that the respondent has accompanied a family member to a distant place. The term education indicates the total educational achievement of the respondent with values as indicated in Table 5.5.

The self-reported climate and environmental variables, drought, flooding, riverbank erosion and cyclone denote answers to the survey questions about experiences of these hazards during the respondent's stay at a particular place. Each variable is constant for the entire duration of a respondent's stay at one place in the dataset. So if a person stays in his village of origin for 20 years and he or she reports experiencing a cyclone, this experience is considered to be constant over the 20 years. It does not mean that the respondent has experienced cyclones over such a long period – but just that he or she experienced the hazards during the stay at the particular place. Capturing specific occurrences and their lasting impacts across a long time span is difficult in a retrospective survey that depends on people's memories. However, the variable based on observed data, the rainfall anomaly, varies from year to year.

Logistic regressions incorporating all these variables – individually, one by one, as well as adding one after another in various combinations – shows the direction (positive or negative) and strength of the relationship between migration and climate and environment hazard experience and risk concern.

5.4 Analysis and findings

5.4.1 Main logit model – first migration outside the district

The main logit model A analyses the first migration outside the district against a set of socio-economic, climatic and environmental variables. Table 5.7 describes the odds ratios in a logit model based on dummies for each year, as is done in the literature (Henry et al. 2004). The coefficients denote the values for the logistic

Table 5.5 **Socio-economic variables** (and their assigned values in the logit models)

Educational achievement	
1	No Schooling
2	Junior School Cert.
3	Secondary School Cert.
4	Higher Secondary Cert.
5	Degree
Poverty level (*Availability of finances at the place of origin*)	
1	Always sufficient
2	Just sufficient
3	Often insufficient
Assets	
1	1–2 assets
2	3–4 assets
3	3–5 or more assets
Networks	
0–10	Number of friends and relatives outside the district of origin

Table 5.6 **Climatic and environmental variables** (and their assigned values in the logit models)

Negative anomaly in rainfall	
0	No negative anomaly
1	Annual rain below 1 standard deviation (for 30 years observation)
2	Rain below 2 standard deviation
Positive anomaly in rainfall	
0	No positive anomaly
1	Annual rain above 1 standard deviation (for 30 years observation)
2	Rain above 2 standard deviation
Normal rain	
0	Positive or negative anomaly
1	Normal rain
Drought	
0	No experience of drought at the place of origin
1	Experienced drought at the place of origin
Flooding	
0	No experience of flooding at the place of origin
1	Experienced flooding at the place of origin
Riverbank erosion	
0	No experience of riverbank erosion at the place of origin
1	Experienced riverbank erosion at the place of origin
Cyclone	
0	No experience of cyclone at the place of origin
1	Experienced cyclone at the place of origin

Table 5.7 Logit Model A: climatic and environmental variables that influence migration

Model no.	*1*	*2*	*3*	*4*	*5*	*6*	*7*	*8*
	migration	*migration*	*migration*	*migration*	*migration*	*migration*	*migration*	*migration*
Socio-economic variables								
Age at migration	−0.141***	−0.157***	−0.141***	−.142***	−.144***	−.160***	−.160***	−.160***
	(−11.92)	(−11.67)	(−11.88)	(−11.77)	(−11.89)	(−11.83)	(−11.83)	(−11.83)
Level of education	0.213**	237**	0.204*	0.220**	0.222**	0.207*	0.212**	0.214**
	(2.60)	(2.98)	(2.47)	(2.64)	(2.71)	(2.50)	(2.58)	(2.59)
Poverty while at the place of origin	0.313*	0.246	0.316*	0.352*	0.380*	0.327*	0.314	0.309
	(1.98)	(1.52)	(2.00)	(2.25)	(2.46)	(2.05)	(1.93)	(1.90)
Assets owned at the place of origin	0.643***	0.636***	0.666***	0.711***	0.725***	0.794***	0.761***	0.762***
	(4.13)	(3.98)	(4.29)	(4.56)	(4.66)	(4.84)	(4.60)	(4.60)
Friends and relatives outside the district of origin	0.508***	0.339***	0.508***	0.511***	0.485***	0.327***	0.333***	0.329***
	(6.00)	(3.95)	(5.92)	(6.00)	(5.81)	(3.84)	(3.92)	(3.88)
Hazard experiences at the place of origin								
Droughts	0.260	0.719***	0.184				0.259	0.253
	(1.24)	(3.38)	(−0.86)				(0.98)	(0.95)
Negative anomaly in rainfall (*station data*)		−0.104				−0.138		
		(−0.74)				(−1.06)		
Flooding			−0.182	−0.0946	−0.319	−0.562**	−0.508*	−0.510*
			(−0.88)	(−0.48)	(−1.55)	(−2.88)	(−2.51)	(−2.52)
Riverbank erosion				−0.509**	−0.521**	−0.579**	-460	−0.459
				(−2.82)	(−2.92)	(−3.16)	(−1.93)	(−1.89)
Cyclones					0.548**	−0.0281	0.0256	−0.00203
					(−2.85)	(−0.14)	(0.12)	(−0.01)
Positive anomaly in rainfall (*station data*)							0.137*	
							(2.31)	
Normal rainfall (*station data*)								−0.439***
								(−4.46)
_cons	2.133***	2.628***	2.285***	2.439***	2.445***	3.469***	3.238***	3.306***
	(3.57)	(4.38)	(−3.67)	(3.75)	(3.75)	(5.33)	(4.95)	(5.04)
N								
adj. R-sq	28066	22437	28066	22437	28066	22437	22437	22437

(t statistics in parentheses * p<0.05, ** p<0.01, *** p<0.001)

regression equation for predicting the dependent variable from the independent variable. The observations are in log-odds units. The t statistics associated with the coefficients are given in brackets.

The table shows that the older the respondent, the lesser he or she is likely to migrate out of the district. In all the regression models (1–8) that incorporate various climatic and environmental hazards, age is a highly significant ($p<0.001$) factor that negatively influences migration. Education appears to be a very significant ($p<0.01$) factor that drives migration in models 1–2, 4–5 and 7–8; it is still significant ($p<0.05$) in the remaining two models. Poverty levels also appear to positively influence migration significantly in models 1 and 3–6. The total number of assets appear to have an even more significant positive influence on migration in all the models – so does the number of networks or family and friends outside the district of origin. These two variables show highly significant results across all the models.

Among the climate- and environment-related variables, experience of drought appears to be a highly significant variable that drives migration in model 2 that incorporates negative anomaly in rainfall. In itself, negative anomaly in rainfall does not become a highly significant variable, though positive anomaly in rainfall tends to drive migration in a significant relationship (model 7). However, normal rainfall has a highly significant negative influence on migration, as model 8 shows. Experience of flood becomes a very significant variable in model 6 and significant in models 7–8, negatively influencing migration. Erosion of riverbanks also appears to influence migration negatively in a very significant manner, as models 4–6 show. Cyclones appear to have a very significant positive influence on migration outside the district that incorporates floods and riverbank erosion, as model 5 shows. However, this influence appears to diminish in other models. Riverbank erosion appears to negatively influence migration in all the models.

5.4.2 Graphic representation of the main logit model using kernel density estimate graphs

It may be noted that the way age at migration is distributed in the dataset is not as if it is in a straight-line graph. Besides, there are district-wise variations in migration patterns. Kernel density estimations show this as a more nuanced picture. Kernel density is a non-parametric density estimator that takes into account all the data points in analysis time to reach an estimate. It works as an interpolation technique for showing individual points in time (Silverman 1986). Here the distribution is estimated by summing the individual kernel functions at different points in time to produce a smooth graph, each point contributing equally to a smoothing probability density graph (Levine 2010).

As the graphs show, the highest probability of migrations, it appears, is when the migrant is in his or her 30s. Though this can be interpreted as the peak probability period, there are also district-wise differences in this trend, as Figure 5.2 shows, and they include the probability of migration at an earlier or later period. There are also differences in the probability of men and women migrating as well. Figure 5.2 and 5.3 shows this.

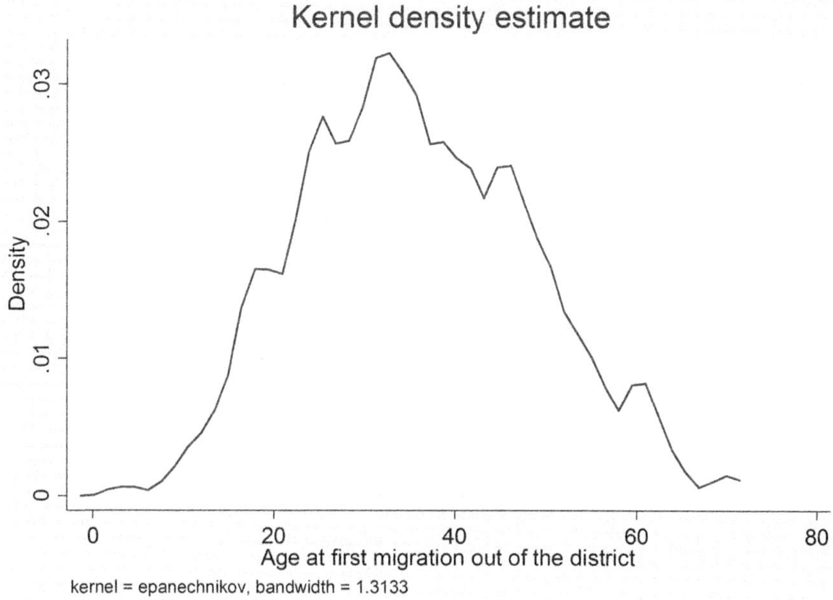

Figure 5.1 Kernel density graph showing the probability of first migration by age

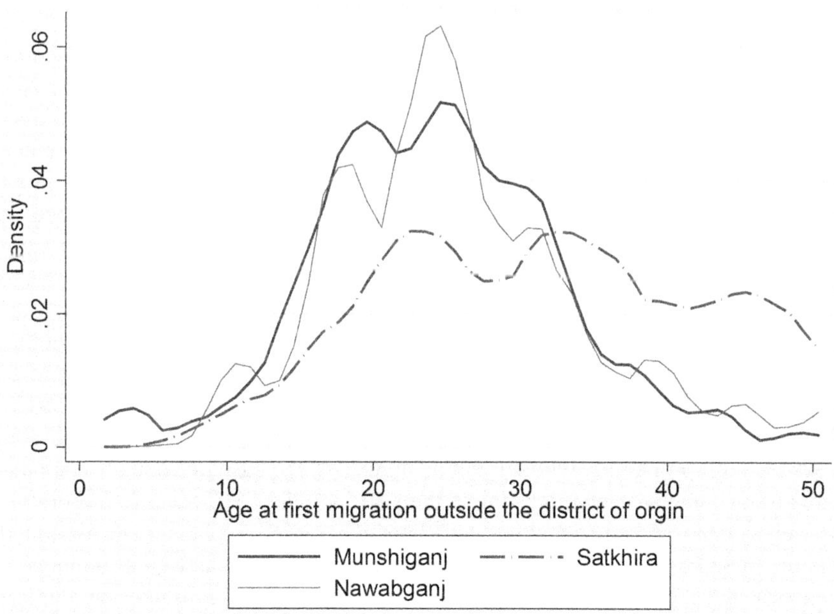

Figure 5.2 Kernel density graph showing first migration by age and district

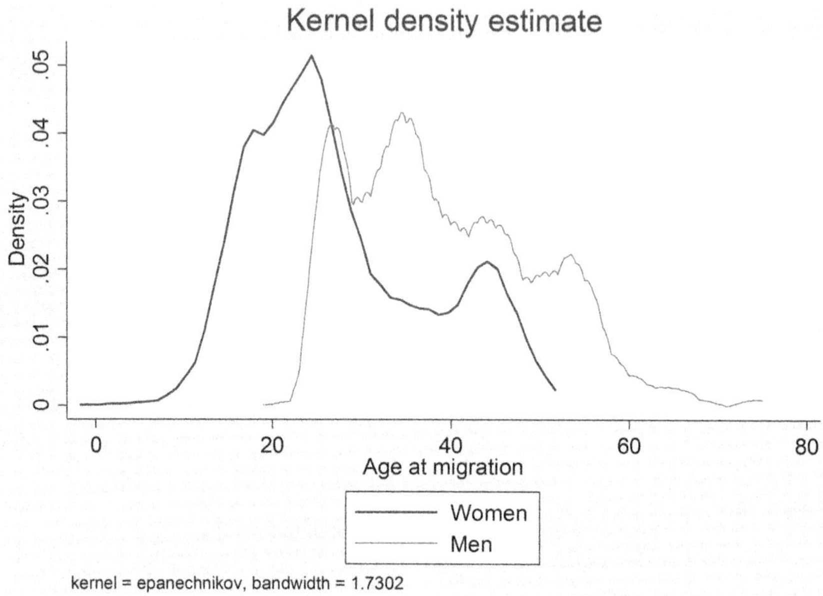

Figure 5.3 Kernel density graph showing the age of women and men migrating

While men and women are the most likely to migrate in their early 20s, there seems to be higher probability of women migrating at an earlier age compared with men, and they tend to be more likely to stop migrating in their late 20s. However, migration of women in their late teens is a phenomenon noticed only in the drought-prone Nawabganj district, as Figure 5.4 shows. This could be due to a linkage between drought and migration (Findlay and Geddes 2011). Logit Model A (Table 5.7) shows that droughts have a highly significant influence on inter-district migration. Across the study areas, one possible explanation for early migration of women is marriage, because women are usually expected to move to their husband's place as the tradition in most parts of Bangladesh mandates. Marriage appears to influence women's migration, as correlation tests show. Another possibility is labour migration by young women to cities. There is is trend of young women working in garment factories before marriage (Amin et al. 1998).

The probability of men's migration tends to peak in their early 20s, and mid-30s. There is high probability (though not as high as in the case of younger respondents) of men and women migrating in their 40s and later. However, as Figure 5.2 as well as tables 7.4 and 5.5 show, such later age migration phenomena are found mainly in Satkhira, the cyclone prone district. The literature as well as qualitative analysis shows evidence for inter-district migration of those affected by the last two major cyclones, namely Sidr (2007) and Aila (2009).

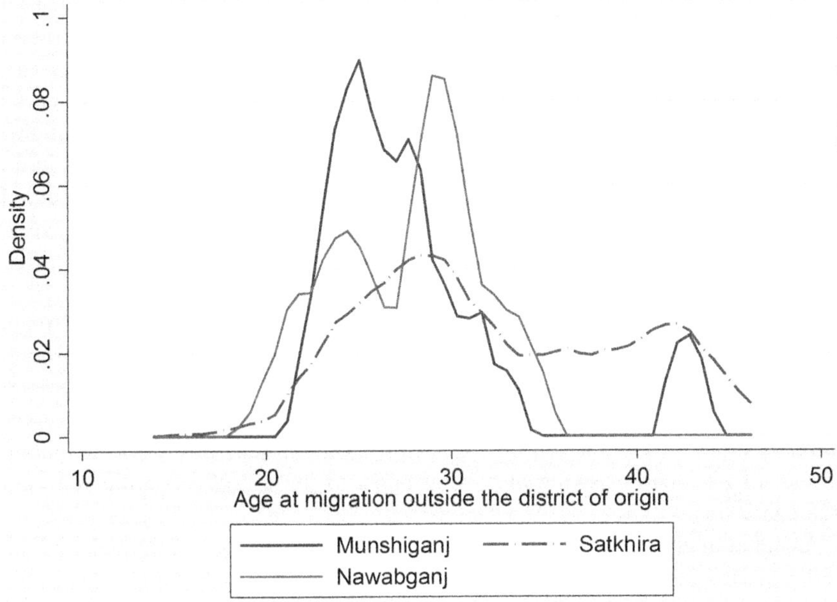

Figure 5.4 Kernel density graph showing the age of women migrating, by district

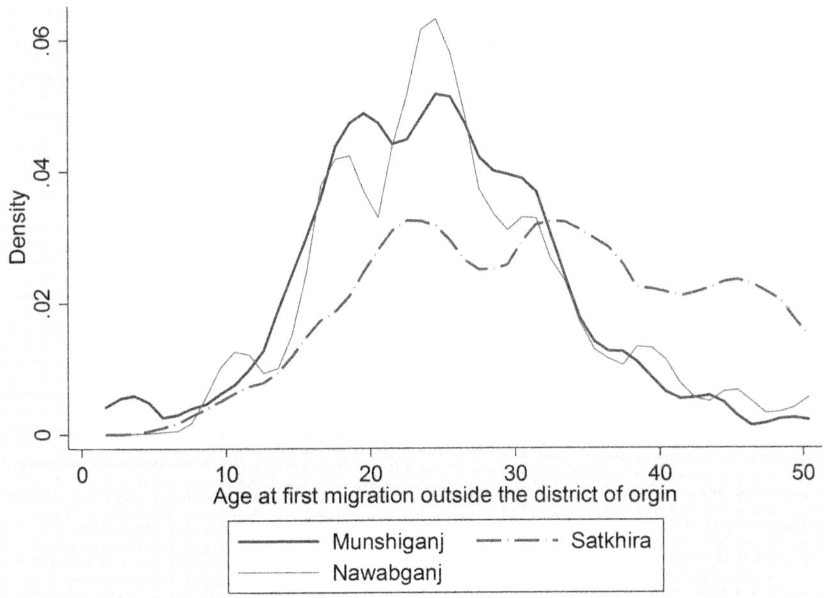

Figure 5.5 Kernel density graph showing the age of men migrating, by district

The kernel density estimate graphs show that there are differences along district lines in the migration patterns of men and women. The probability of migration peaking at early or late 20s is largely a phenomenon in Munshiganj and Nawabganj; whereas migration from Satkhira appears to continue in a smoother curve as men and women get older. This trends points at the possibility that migration from Munshiganj and Nawabganj involves mostly young people driven more by income needs, whereas those who leave Satkhira include people more settled in life, possibly leaving the district after the impacts of recent cyclones. Census trends show trends (BBS 2012) of slow population growth in the coastal areas, suggesting migration away from the coasts (Marshall and Rahman 2013).

Analysis of census data show that coastal divisions of Barisal and Khulna (that includes Satkhira district) perform very weakly, posting population growth well below the national average, suggesting large-scale outmigration, especially during 2001–2011 (Marshall and Rahman 2013). While environmental stresses and shocks are a possible reason for such migration, generally weak economic growth of the coastal region also contributes to these outflows (Marshall and Rahman 2013).

As noted in chapter 3, studies show that migrants from the coastal belt and the northern Monga-effected districts (that includes Nawabganj) comprise a sizeable share of slum dwellers in Dhaka. This has been the trend even before the 2007 and 2009 cyclone. As a Centre for Urban Studies survey done in 2005 noted, people from coastal areas accounted for 31.9 per cent and Monga-effected northern districts 4.6 per cent of Dhaka slum dwellers (Marshal and Rahman 2013). The northern district does not show such dramatic outmigration as the coastal districts (Marshall and Rahman 2013).). Possibly a part of the migration comprises seasonal migration that is not captured in this analysis. It appears that cyclone events, their lingering impacts such as soil salinity, the perception of future cyclone risk and the general economic backwardness of the coastal region have together shaped the migration decisions of people in Satkhira.

5.4.3 Graphic representation of the main logit model using Kaplan-Meir graphs

The migration trends are further illustrated in the separate models for men and women attached later in this chapter. In a Kaplan-Meier survival analysis graph, each subject has three variables – the serial time, status at the end of the serial time (event occurrence or censoring) and the group they are in. The horizontal lines along the serial time indicate the survival duration. The interval ends at the occurrence of the event – in this case migration. The vertical distances between horizontal lines show the change in cumulative probability as the curve moves along the event time. Kaplan-Meier curves are non-continuous, so they do not look smooth, unlike kernel density estimates, but are step-wise estimates (Rich et al. 2010). In the examples below, the cumulative probability of surviving (not migrating out of the district of origin, even when changing houses within the district) during a given time is seen on the Y-axis. The steepness of the curve is determined by the survival durations shown as length of horizontal lines.

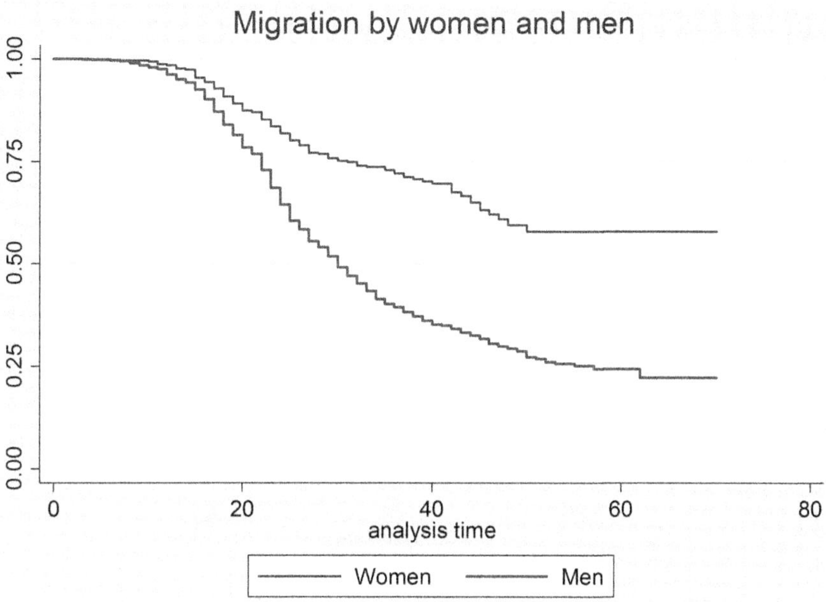

Figure 5.6 Kaplan-Meir survival graph showing migration of women and men

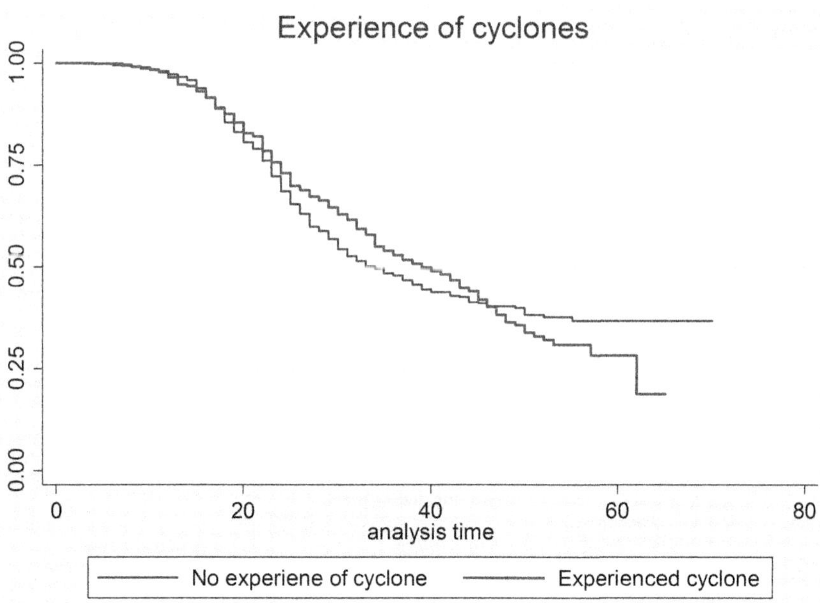

Figure 5.7 Kaplan-Meir survival graph showing migration with relation to cyclones

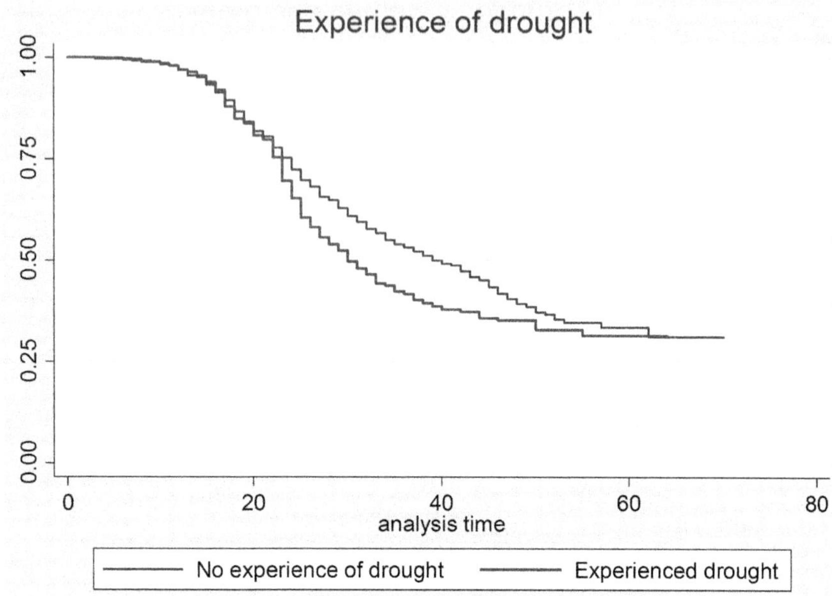

Figure 5.8 Kaplan-Meir survival graph showing migration with relation to droughts

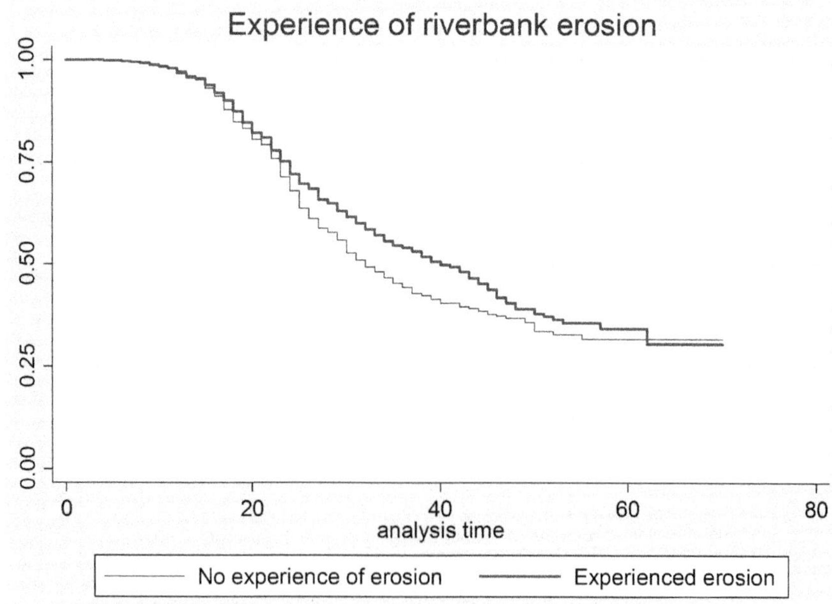

Figure 5.9 Kaplan-Meir survival graph showing migration with relation to riverbank erosion

Men tend to have lower levels of survival (that means migrate more) than women. As for climatic and environmental variables, while droughts appear to drive migration, and erosion discourages migration, cyclones show a mixed influence. While in early years it drives migration, the trend reverses from the 20th year onwards, but picks up again after the early 40s. Such a mixed pattern is shown clearly in the kernel density graph showing migration patterns of men in the cyclone affected district Satkhira.

5.4.5 Supplementary models showing differences in first migration among men and women

As discussed above, socio-economic and environmental variables work differently in migration of men and women. Though the whole picture of migration requires us to take into account migration by all adults, it may be worthwhile examining how the influences being analysed in this chapter work exclusively on men, women and together in a combined manner. The following models show analysis disaggregated along gender lines.

The gender-disaggregated models Logit Models A-m and A-w show differences on how socio-economic, climatic and environmental variables influence migration. The strongest common influences on migration in the men's and women's models appear to be age, showing a highly significant negative correlation across all models, and social networks, showing high significance as a positive influence in all women's models and the first five (1–5) men's models (1–5). Riverbank erosion comes across as a strong negative influence in three of the men's models (1–3), attaining more influence than in the main model. The experience of cyclones also appears to be a highly significant influence in only one women's model (as in the main model, but with added significance). Assets, a highly significant positive influence in the main model, appears to have lost its significance in the men's and women's models. However, poverty retains its significance only in the first three women's models (1–3) at the same level as in the main model (1–6).

5.4.6 Supplementary logit model – moving house for the first time

As most of the respondents (about 87 per cent) have moved house at least once, and most of their movements (as much as 74 per cent) were within the district, it may be worthwhile to look at what drives such shifts, and how they are related.

> Similar to the main migration model, the model of house shifts (largely within the district) shows the older the respondent, the less likely they are to move house. In all the regression models (1–8), age at the time of moving house remains a highly significant factor. The number of social networks (friend and family), however, is a highly significant positive influence across the first five models (1–5) as in the main model, but loses the level of significance in the rest (5–6).
>
> Cyclones appear to have a highly significant (models 6–8) negative influence over house shifts – this is contrary to what has been seen in long-distance

Table 5.8 Logit Model A-m: first migration of men out of the district of origin

Model no	*1*	*2*	*3*	*4*	*5*	*6*	*7*	*8*
	migration	*migration*	*migration*	*migration*	*migration*	*migration*	*migration*	*migration*
Socio-economic & demographic variables								
Age at	−0.157***	−0.166***	−0.157***	−0.157***	−0.158***	−0.165***	−0.165***	−0.165***
migration	(−11.61)	(−11.09)	(−11.66)	(−11.79)	(−11.87)	(−11.25)	(−11.11)	(−11.10)
Level of education	0.0593	0.0821	0.0699	0.104	0.102	0.0963	0.0974	0.0980
	(0.58)	(0.82)	(0.69)	(0.99)	(0.97)	(0.92)	(0.94)	(0.94)
Poverty while at the place of origin	0.215	0.224	0.213	0.246	0.281	0.249	0.250	0.247
	(1.05)	(1.11)	(1.04)	(1.19)	(1.39)	(1.21)	(1.22)	(1.20)
Assets owned at the place of origin	0.333	0.342	0.306	0.306	0.280	0.354	0.354	0.355
	(1.62)	(1.78)	(1.49)	(1.48)	(1.36)	(1.73)	(1.73)	(1.74)
Friends & relatives outside the	0.412***	0.263*	0.413***	0.422***	0.404***	0.270*	0.272*	0.268*
district of origin	(3.32)	(2.26)	(3.37)	(3.46)	(3.33)	(2.32)	(2.33)	(2.30)
Hazard experiences at the place of origin								
Droughts	0.404	0.725**	0.482*				0.183	0.183
	(1.67)	(2.86)	(2.00)				(0.53)	(0.53)
Negative anomaly in rainfall		−0.151				−0.145		
(*station data*)		(−0.84)				(−0.86)		
Flooding			0.180	0.318	0.123	−0.00115	0.0279	0.02
			(0.70)	(1.23)	(0.44)	(−0.00)	(0.11)	(0.10)
Riverbank erosion				−0.865***	−0.857***	−0.841***	−0.731*	−0.721*
				(−3.49)	(−3.47)	(−3.41)	(−2.11)	(−2.08)
Cyclones					0.422	−0.0928	−0.0635	−0.0831
					(1.57)	(−0.33)	(−0.23)	(−0.30)
Positive anomaly in rainfall							0.100	
(*station data*)								
							(1.50)	
Normal rainfall (*station data*)								−0.376**
								(−3.16)
_cons	4.118***	4.266***	3.987***	4.429***	4.415***	4.879***	4.693***	4.746***
	(5.58)	(5.59)	(5.48)	(5.88)	(5.90)	(6.07)	(5.88)	(5.94)
N	17562	13613	17562	17562	17562	13613	13613	13613
adj. R-sq								

(t statistics in parentheses * p<0.05, ** p<0.01, *** p<0.001)

Table 5.9 Logit Model A-w: first migration of women out of the district of origin

Model no.	*1* *migration*	*2* *migration*	*3* *migration*	*4* *migration*	*5* *migration*	*6* *migration*	*7* *migration*	*8* *migration*
Socio-economic variables								
Age at migration	−0.130***	−0.204***	−0.130***	−0.126***	−0.134***	−0.210***	−0.211***	−0.210***
	(−5.47)	(−4.22)	(−5.47)	(−5.14)	(−5.78)	(−4.42)	(−4.37)	(−4.37)
Level of education	0.224	−0.0448	0.224	0.223	0.286	−0.0851	−0.0796	−0.0792
	(0.94)	(−0.31)	(0.94)	(0.96)	(1.34)	(−0.53)	(−0.48)	(−0.48)
Poverty while at the place of origin	0.641*	0.337	0.640*	0.664*	0.739*	0.534	0.530	0.526
	(2.26)	(1.24)	(2.24)	(2.31)	(2.45)	(1.96)	(1.92)	(1.90)
Assets owned at the place of origin	0.0706	−0.189	0.0660	0.0879	0.368	0.298	0.249	0.248
	(0.19)	(−0.46)	(0.18)	(0.26)	(1.12)	(0.83)	(0.59)	(0.58)
Friends and relative outside the district of origin	0.818***	0.722***	0.818***	0.836***	0.792***	0.682***	0.685***	0.684***
	(5.84)	(4.12)	(5.87)	(6.13)	(5.84)	(4.17)	(4.19)	(4.19)
Hazard experiences at the place of origin								
Droughts	−0.415	0.730	−0.404				0.175	0.161
	−1.12)	(1.70)	(−1.07)				(0.36)	(0.33)
Negative anomaly in rainfall (*station data*)		−0.223*				−0.177		
		(−2.02)				(−1.92)		
Flooding			0.0253	0.378	−0.114	−0.652	−0.601	−0.603
			(0.07)	(0.94)	(−0.27)	(−1.52)	(−1.40)	(−1.40)
Riverbank erosion				−0.635	−0.742*	−1.007*	−0.962*	−0.952*
				(−1.85)	(−2.15)	(−2.31)	(−1.99)	(−1.97)
Cyclones					1.311***	0.278	0.327	0.306
					(3.70)	(0.65)	(0.75)	(0.70)
Positive anomaly in rainfall (*station data*)							0.00810	
							(0.06)	
Normal rainfall (*station data*)								−0.553*
								(−1.98)
_cons	−0.577	2.391	−0.595	−0.869	−1.135	3.183	3.051	3.083
	(−0.47)	(1.35)	(−0.48)	(−0.64)	(−0.84)	(1.70)	(1.66)	(1.68)
N								
adj. R-sq	10504	8824	10504	10504	10504	8824	8824	8824

(t statistics in parentheses * p<0.05, ** p<0.01, *** p<0.001)

Table 5.10 Logit model B: first instance of moving house

Model no	*1*	*2*	*3*	*4*	*5*	*6*	*7*	
	move	*move*	*move*	*move*	*move*	*move*	*move*	
Socio-economic & demographic variables								
Age at	−0.152***	−0.169***	−0.151***	−0.154***	−0.151***	−0.169***	−0.168***	−0.168***
migration	(−14.53)	(−13.69)	(−14.39)	(−14.15)	(−13.92)	(−13.70)	(−13.44)	(−13.47)
Level of education	−0.230*	−0.214*	−0.231*	−0.238*	−0.257*	−0.269*	−0.269*	−0.267*
	(−2.36)	(−2.14)	(−2.35)	(−2.41)	(−2.57)	(−2.52)	(−2.52)	(−2.51)
Poverty while at the place of origin	0.309	0.277	0.343	0.340	0.326	0.218	0.230	0.230
	(1.57)	(1.37)	(1.73)	(1.73)	(1.64)	(1.03)	(1.08)	(1.08)
Assets owned at the place of origin	0.106	0.207	0.161	0.145	0.186	0.35	0.348	0.348
	(0.57)	−1.09	(0.86)	(0.77)	−0.99	−1.77	−1.75	(1.75)
Friends & relatives outside the	0.502***	0.347*	0.515***	0.528***	0.588***	0.445**	0.445**	0.444**
district of origin	(3.73)	(−2.34)	(3.84)	(3.99)	(−4.3)	(−2.83)	(−2.84)	(−2.83)
Hazard experiences at the place of origin								
Droughts	−0.494*	−0.0903	−0.771***				−0.222	−0.221
	(−2.32)	(−0.40)	(−3.32)				(−0.63)	(−0.62)
Negative anomaly in rainfall		0.124***				0.0598*		
(*station data*)		(−4.27)				(−2.33)		
Flooding			−0.657**	−0.535*	−0.0979	−0.292	−0.324	−0.324
			(−2.60)	(−2.23)	(−0.34)	(−1.01)	(−1.12)	(−1.12)
Riverbank erosion				0.600**	0.734**	0.722**	0.578	0.59
				(2.65)	(3.12)	(2.93)	(1.72)	(1.76)
Cyclones					−0.809**	−1.056***	−1.048***	−1.072***
					(−2.86)	(−3.53)	(−3.41)	(−3.48)
Positive anomaly in rainfall							0.258***	
(*station data*)							(5.83)	
Normal rainfall (*station data*)								0.589***
								(4.63)
_cons	5.098***	5.256***	5.485***	4.913***	4.794***	5.510***	5.579***	5.633***
	(7.41)	(7.02)	(7.82)	(7.28)	(7.17)	(7.01)	(6.92)	(6.99)
N	31394	20354	31394	31394	31394	20354	20354	20354
adj. R-sq								

(t statistics in parentheses * p<0.05, ** p<0.01, *** p<0.001)

migration. Drought when combined with negative rainfall anomaly becomes a highly significant factor that negatively influences house shift in one of the models – it is the opposite of what is seen in one of the main models. Negative rainfall anomaly tends to drive movement, with a highly significant influence in one of the models (model 2). Experience of floods tends to discourage movement, with a very significant (model 3) or significant (model 4) influence.

5.5 Discussion

5.5.1 Lessons from the main model

The results suggest that climate- and environment-related hazards influence migration in non-linear and indirect ways. They sometimes do not necessarily drive migration; on the contrary, sometimes they appear to hinder migration. Most of the time they work in combination with socio-economic variables that can be considered primary drivers of migration – they include age of the migrant, poverty levels, assets, networks and education. The logistic regression models described above show varying levels of such influences and how they change in the presence of various climate- and environment-related variables – and the choice and distance of migration destination.

The observations in general in the main Logit Model A appear to be in line with literature mentioned in in chapters 2 and 3. The model shows that migration out of the district is an age-dependent, poverty-driven activity, strongly supported by social networks. However, people who migrate are those who have some assets that they could use to cover the costs involved, not necessarily the poorest in the village. Chronic exposure to drought makes people highly prone to inter-district migration, possibly in search of city-based livelihoods. It may be noted that the top destinations of the inter-district migrants are big urban centres, namely, Dhaka, Khulna, Chittagong, Rajshahi and Jessore. However, normal rain appears to discourage migration in a highly significant manner. It could be argued that especially in chronically drought-affected places, normal rainfall comes as a blessing that offers the prospect of good crops, encouraging people to stay back. Floods and riverbank erosion tend to discourage migration. This could be due to a sudden shock that reduces resources to migration. As such, the literature suggests that flood-induced movements tend to be short-term and short-distance, and erosion-related displacement tends to involve short distances, though it could be for a long duration. However, this aspect needs deeper probing to understand the underlying dynamics. Cyclones tend to drive migration outside the district, as the literature suggests.

In short, migration appears to be driven by the need for better income, and people with better assets, education and networks are better placed to migrate outside the district. Positive environmental events such as a normal rain or shocks such as flood or riverbank erosion could discourage migration. At the same time, more dramatic – and potentially more devastative events – such as cyclones could drive migration outside the district.

5.5.2 Differences along gender lines

The different trends observed show that men's and women's migration patterns show that influences of socio-economic, climatic and environmental factors differ across gender lines. Among the socio-economic variables, men's and women's models retain the high significance of age and social networks that influence migration in negative and positive manners, respectively. While poverty and assets retain the same positive relationship to migration, they have lost their significance in men. Poverty retains its significance as in the main logit models (models 1, 3–5, but not in 6).

The results may be interpreted as showing that while it is young men and women with education who are more prone to migrate, income needs drive migration of women more significantly compared with men. The only climatic hazard that has a (highly) significant positive influence on women's migration is cyclone. It appears that women and families seem to be moving out of cyclone-affected areas, as figures 7.2, 7.4 and 7.5 (that show general, men's and women's migration from the cyclone-prone Satkhira) show. This aspect is further discussed with comparisons with the literature in chapter 9. Meanwhile riverbank erosion, normal rainfall and even a negative anomaly in rainfall appear to have a negative influence on women's (long-distance) migration. The relationship of rainfall anomalies with migration has been discussed in sub-section 7.4.6. An inverse relationship between negative rainfall anomaly and migration could mean that families or women do not move out in a year of bad rain.

Some of the variables gain significance only in the main model. Assets appear not to have significant influence on men's or women's migration, as gender disaggregated models show. However, assets positively influence migration in a highly significant manner, as the main model shows. It looks like a case of sum total significance becoming greater than the significance shown by the two components, namely women's and men's migration. The negative influence of flooding is another variable that has lost its significance in the separate models. The reason could be almost halving of the sample size. When the sample is divided, the regressions tend to lose statistical significance. Estimating separate models for each group can result in loss of statistical power, i.e. it may be less likely to reject the null hypothesis (Williams 2015). It just means that the analysis loses degrees of freedom with reduced sample size in the regression, not that any of the analyses were incorrect.

5.5.3 Comparing with moving house

In general, Logit Model B shows that a set of socio-economic factors as well as climatic and environmental factors highly influence the first instance of moving house. However, these influences are somewhat different from those that influence the first migration out of the district. The highly significant negative influence of age shows a trend similar to that of Logit Model A. However, changing house appears to be driven by poverty, but not aided by ownership of more assets, as inter-district migration is. Networks appear to be highly significant – as much as in inter-district migration – in determining shifting houses. This predominant influence of poverty and networks in the first instance of moving house – that can be counted as short-distance internal migration – is in line with the literature, as explained above.

Experience of droughts and floods appears to discourage moving house to some extent, and cyclones tend to have an even stronger hindering effect with consistently very/highly significant negative relationship with movement. It may be inferred that people are probably incapacitated by exposure to hazards to find or buy a new house and move to a better place, in a sense, trapping them in their place, as the literature shows. One hazard that appears to drive a house shift is riverbank erosion. Movements following riverbank and coastal erosion tend to be short-distance, as the literature and the qualitative analysis shows. While individual deficit rain years tend to drive migration, excess rain or even normal rain appear to encourage a house shift – possibly for different reasons. It could be a case of the prospects of a bumper crop following good rains that aids migration or crop loss in a flood that requires migration for work or moving away from a flooded place.

5.6 Conclusion

The evidence furnished in this chapter suggests that it is predominantly income needs that drive migration, as shown clearly in the chapter 4. People with income needs tend to migrate more. Though it may appear to be contradictory, among those who are in need of more income, it is people with more assets that actually migrate. It shows that though migration is driven by poverty and income needs, the poorest, those without any assets, are often unable to migrate outside the district. The youth – rather respondents in their 20s – migrate more than others do. Migrations are boosted by the educational attainment of the migrant and boosted by social networks.

Among the climatic and environmental variables, experiences of drought and cyclone tend to positively influence migration outside the district – and riverbank and coastal erosion negatively. Rainfall uncertainties show different influences – with negative anomaly in rainfall showing no significance, but positive anomaly driving migration and normal rainfall positively and influencing migration in a highly significant manner. Experience of flood becomes a very significant or significant factor that negatively influences migration. Migration of women tends to take a different pattern compared with that of men. Inter-district differences in migration are prominent, especially in the migration of middle-aged people from the cyclone-prone Satkhira and women in their late teens from the drought-prone Nawabganj.

The first instance of moving house is influenced by some of the drivers of inter-district migration, especially youth, poverty (not assets as in migration) and networks. Experience of droughts and floods appears to discourage moving house, and cyclones tend to have an even stronger hindering influence. Riverbank erosion tends to drive local movements. Excess rain or even normal rain appears to encourage a house shift. The general trends revealed in the regressions show that in contrast with many studies in the older climate change and migration literature (for instance, Ticklll 1989; Homer-Dixon and Percival 1996), climatic and environmental hazards do not always drive migration – instead they sometimes tend to hinder migration, as the models reveal. It is more of a story of mixed and differential influences.

6 Removing roadblocks

Policy challenges in making migration an adaptation strategy

A textual analysis of current and recent policies concerning climate change, development and poverty alleviation, and disaster management shows that the economic and adaptive roles of internal migration are usually not included in policy framing. In contrast, this chapter argues that if migration works as a positive step towards adaptation, a range of policies could encourage and facilitate it, rather than ignoring or inhibiting it. The key challenge, however, is to align the policies with Bangladesh's development and existing and projected migration patterns. The chapter looks at these challenges and updates recent efforts on this front in not only Bangladesh, but also other countries that face similar climatic and environmental threats.[1]

6.1 Sharing lessons

On a balmy August day in 2016, far away from the Gangetic Delta at Lomé, the capital of Togo on the coast of the Atlantic, a group of government officials listened keenly to their counterpart from Bangladesh. Moinul Islam Moin, a young officer from the Planning Commission of Bangladesh, told them about his government's efforts to mainstream climate change into different areas of planning and to address climate-related migration and displacement. A key document that he presented was the National Strategy on the Management of Disaster and Climate Induced Displacement, a new paper that is now part of Bangladesh's long-term planning process (Siddiqui et al. 2015). An outcome of a set of studies that looked closely at climate-related migration in Bangladesh, notably the CDKN-funded project that is the basis of this book, the strategy paper takes a rights-based approach. It critiques the traditional relief-oriented approach to displacement issues and favours a set of more proactive and comprehensive responses. Such a new approach incorporates the government's commitment to disaster risk reduction and climate change adaptation strategies under the Sendai Framework for DRR (2015–2030) and a set of national laws. The Sendai Framework set out the roadmap for responding to disasters worldwide by keeping the rights of the disaster-affected people at the forefront.

In a country that faces frequent droughts, floods and migration within and across its borders, Togolese officials are trying to come to grips with this dynamic situation. The Bangladesh experience shared as part of the training programme offered

1 This chapter draws substantially in research and content from Martin et al (2017).

by the International Centre for Migration Policy Development (ICMPD), Brussels, became useful for a country that shares many of the development challenges that Bangladesh faces. In Togo with streets buzzing with markets, vendors and migrant workers going to work, as well as in the busy, crowded streets of Dhaka, one can see the effect of a growing city-based economy. Migration is very much part of development in these two countries, as elsewhere in the developing world.

This chapter traces the policy processes in Bangladesh that promote or limit migration. It acknowledges that villagers in Bangladesh migrate for better livelihoods, amid climate- and environment-related stresses and shocks. However, these factors need not necessarily influence migration, and sometimes they make migration difficult or impossible. Nonetheless, in the context of a growing, city-centred economy based on urban industries and services that promote urban migration (Muzzini and Aparicio 2013), migration can be considered as an effective adaptation strategy, as the literature suggests (McLeman and Smit 2006; Barnett and Webber 2010; Tacoli 2009; Foresight 2011; ADB 2012; IPCC 2014; for instance). Migration is a way for people to earn better, and cover losses suffered in climate- and environment-related events that are very frequent in the country. Bangladesh has a history of high levels of internal migration, with or without climate change (Afsar 2003; Gardner 2009).

6.2 Migration as an adaptation strategy

In this background of high levels of mobility and climatic stresses and shocks, migration could be seen as one of the many adjustments that people make in response to actual or expected climatic stimuli or their effects. If that is the case, as per the IPCC definition of adaptation (Parry et al. 2007), migration can be considered an adaptive strategy. During climate- and environment-related stresses and shocks, including water shortage, cyclone, floods and coastal/delta erosion (Adams et al. 2011a), migration also contributes to adaptive capacity of people by giving them better access to resources, livelihoods, markets and social networks (Gerlitz et al. 2014).

Still, often the poorest and the most vulnerable people are unable to migrate out of environments exposed to hazards, as the quantitative analysis shows. Experience of hazards appears to limit their capacity to migrate. Still the major policy concern in Bangladesh is not how to mobilise migration as a form of adaptation, but whether climate change will increase climatic variability and/or the frequency of extreme events, thus adding to migration flows (Black et al. 2011b). Indeed, a textual analysis of current and recent policies concerning climate change, development and poverty alleviation, and disaster management shows that the economic and adaptive roles of internal migration are often usually not included in policy framing.

In contrast, this chapter argues that if migration works as a positive step towards adaptation, a range of policies could encourage and facilitate it, rather than ignoring or inhibiting it. The key challenge, however, is to align the policies with Bangladesh's development and existing and projected migration patterns.

The first section briefly explains the methods used in this research. Based on the qualitative and quantitative findings, the following section explains how migration can be viewed as an effective adaptive strategy. The third section analyses how different government policies in Bangladesh – concerning climate change, development and disaster risk reduction – view and deal with migration in the context of climate change and development and consider its potential as a climate change adaptation measure. Based on this examination, the paper suggests that there is considerable scope for policy realignment – to acknowledge, plan and promote migration under appropriate circumstances rather than inhibiting it. Finally, by way of conclusions, it lists out a few possible areas of policy attention and action (Martin et al. 2017).

Interviews and focus groups with villagers in places affected by droughts, floods, riverbank erosion and cyclones show that poor people often have to migrate to recover from losses they have suffered on account of these hazards and disasters. Often people lose their crops, croplands and farming and fishing labour opportunities because of these hazards. The numeric analysis confirms this trend. Migration is largely driven by poverty; still climatic and environmental factors do play a role. That role may not be direct, explicit or acknowledged by the migrants. At the same time, the quantitative analysis shows that people exposed to disasters over a long term and those who are the poorest, especially without any assets, are often unable to move out of their place. That is because migration costs money.

Studies show that a trend of increase in extreme weather events and associated disasters in a changing climate (IPCC 2012; World Bank 2010) can alter migration patterns in Bangladesh, especially from the coastal areas that are vulnerable to cyclones and salinity intrusion (World Bank 2010; BBS 2012; Marshall and Rahman 2013). That is possibly because of the impoverishing effects of exposure to these hazards. In our interactions, migrants have reported that they could earn enough during these migrations and offset losses suffered in disasters. Some of them come back and start their lives again, some stay back in the destinations and take their families there and still others stay and send home money regularly so that their family members could continue living in the village with an extra income by way of remittances. Therefore, it can be argued that migration works as an effective adaptation strategy in the face of climatic and environmental stresses and shocks, even though people perceive and report it more as an economic activity. Considering this adaptation potential of migration, it may be worthwhile to consider ways to mainstream it into development policies.

The argument here is that if migration is good for climate adaptation and it contributes to Bangladesh's city-based growth, then policies should encourage and facilitate migration and make it part of the development process. We wanted to test whether policies in Bangladesh acknowledge migration as an effective climate change adaptation strategy.

Public policy is all about government action, across economic, social and political spheres (Minogue 1983). Our study considered inter-relationships among these factors as well as how action and inaction of stakeholders could influence the policy-making process (Keeley and Scoones 1999). We tested whether policies across a broad spectrum acknowledge migration as a climate change adaptation strategy in Bangladesh. We analysed the text of a set of key policy documents in the fields of migration, climate change, development and disaster management. We followed up our text analysis with key informant interviews and consultations with senior government officials and other stakeholders, including academics and NGOs. We looked at how certain ideas, values and interests gain – or lose – currency in governance (Iannantuono and Eyles 1997). We also looked at how certain things were not mentioned in the policy, indicating that there are conflicting or contradictory values (Yanow 1992). We examined how policies are interconnected and how they addressed climate- and environment-related migration; and whether they encourage, discourage or keep quiet about such migration. We probed five sets of policies – those related to development, disaster risk reduction, climate change and migration itself.

6.3 Development and poverty reduction policies

Bangladesh being a developing country, just about coming out of its status as a Least Developed Country, there is a lot of policy emphasis on poverty reduction. As chapter 2 notes, much of the growth in Bangladesh happens in its big cities, especially since the early 1990s. Inevitably it is driven by a labour force coming from rural areas. Therefore, migration in Bangladesh is predominantly rural-to-urban, driven largely by a need for better livelihoods and income, as chapters 4 and 5 show, and the pull factor of city-based industries and services. In this context, inevitably a large share of people coming to cities in search of work should come from places affected by climate- and environment-related hazards. The question here is to what extent policy processes understand, acknowledge and facilitate climate- and environment-related migration. Do urban and rural development plans acknowledge the role of migration? How easy or difficult do they make it for migrants to move to a city to get a job and settle in a migrant destination? These were some of the questions we had in mind while we started analysing the policy documents. The analysis included the country's Sixth Five Year Plan (2011-15) and Outline Perspective Plan (to 2021), as well as its Poverty Reduction Strategy Papers (PRSP) and the progress report on Millennium Development Goals (MDG).

As for development planning, poverty alleviation and social sector development were at the top of the government agenda since the country's independence in 1971. Economic policy over the past two decades especially stressed growth. Bangladesh's long planning roadmap is drawn in the five-year plans. Besides, the government reports on its compliance with international commitments in documents such as the National Strategy for Accelerated Poverty Reduction, which reported its achievements of the Millennium Development Goals. These strategy papers are prepared in consultation with development partners, including the World Bank and the International Monetary Fund. With three-year updates, they describe the country's macroeconomic, structural and social policies for growth and poverty reduction as well as major sources of financing – indeed, the Sixth Five Year Plan followed up on these findings. The first document came out in

2005 (IMF 2005). A key focus of the report (Planning Commission 2005) is the emerging rural-urban continuum, including the "dramatic expansion" of all-weather rural infrastructure developed in the 1980s.

From the point of view of migrants, these rural-urban transition zones are important, as they share the complimentary and often conflicting characteristics of their points of origin and destination (Iaquinta and Drescher 2000: 18). Migrants often end up living in these spaces, as renting or buying space closer to the city centre can be prohibitively expensive. Often rural development and urban planning are inevitably connected with one another, and intervention in one sphere influences the other. As rural-urban transition zones, peri-urban spaces are often affected by some of the most serious urbanization challenges, including a prevalence of slums, degradation of farms and inadequate services, including drinking water, sanitation and environmental services (Dodman 2009).

Sometimes planning regulations and land tenure norms are poorly implemented in such areas, increasing the risks borne by poor and other vulnerable people (Wisner et al. 2004; Tacoli 2006). Often these peri-urban zones are at the heart of the political economy of land tenure systems and the balance of power between different owners and users, and migrations often negotiate this space for livelihoods and living. As chapter 4 notes, migrants often occupy peri-urban spaces and work in the informal economy with little access to civic services or amenities. To a large extent, policies do not look at the habitat issues of the migrants, even while a large part of the country's city-based urban growth depends on migrant labour.

The PRSP document, however, has acknowledged the role that migrant labour plays in poverty reduction, and its relevance to the overall development process. It noted that migration and remittances have emerged as dominant factors, with "migration of varying duration to a variety of destinations both rural and urban as well as near and far is increasingly a critical part of the picture. . . . Initial fears that migration was fuelling an export of poverty from rural to urban areas has now been dispelled by poverty trend statistics; in general, urbanisation appears to have been a force for poverty reduction with urban poverty declining much faster than rural poverty ((Planning Commission 2005: xvi). The document offered a window of opportunity to build on this recognition of the economic role of migration.

However, this looks like a case of missed opportunity. Newer documents appear to somehow ignore this positive note about migration. A new policy strategy paper that came out in 2008 (Planning Commission 2008) and its update prepared after the Awami League came to power in 2008 (Planning Commission 2009) do not take this line on internal migration. The 2009 version of the strategy paper seeks ways to encourage remittances, provide loans to international labour migrants, make processes transparent and promote better labour laws of the employing countries. While largely remaining silent on internal migration, the document addresses challenges in livelihoods, disaster risk reduction, forestry and technology. For the urban poor and slum dwellers, cooperatives, micro credit organisations and health facility improvements have been envisaged in the Poverty Reduction Plan (Planning Commission 2009).

It is the five-year plans that blueprint the development activities in Bangladesh. As part of our research, we analysed the Sixth Plan document that provided a blueprint for Bangladesh's development from 2011 to 2015. The document did underline the importance of migration for the country's development (Planning Commission 2011). It noted that in underdeveloped regions, especially Khulna, Rajshahi and Barisal Divisions, international labour migration could enhance development prospects. It talked about the future prospects of hundreds of thousands of labourers finding jobs abroad. Migration found mention in the document in 15 pages. The problem was that while it promoted international migration in all earnestness, internal migration or its complexities were barely considered. It framed climate-related migration as displacement – something that happens to people, not what people decide to do, as the previous chapters showed. Forced migration was seen as a problem created by climate change, but internal migration is not seen as a potential form of adaptation by people affected by it. The plan did address climate change, its impacts and challenges. It was linked to environment and disaster management. The document even projected that flooding and storm surges could displace 70 million people. The Plan recommended mainstreaming climate change into efforts that address poverty and environment-related issues. Climate change was to find place in project design, budget allocation, project implementation and monitoring processes of the government.

During the Sixth Plan period, the government has integrated climate change policy within country-level policy, planning, budgeting, monitoring and evaluation systems. The Planning Commission's Poverty-Environment-Climate Mainstreaming (PECM) aims to mainstream environment and climate change issues into the overall planning process. Under PECM initiative, a UNEP and UNDP-funded project, GoB conducted a Climate Public Expenditure and Institutional Review (CPEIR) in 2012 and recommended the setting up of a climate fiscal framework (CFF). CFF identifies expenditure as well as revenue or finance aspects of climate funds and encourages an open and accountable Climate Fiscal Policy (CFP) (Finance Division 2014). CFF recommends ways to equip budget management centres and climate change cells within line ministries as part of institutional reforms for more effective climate financing and implementation of climate change responses.). In short, CFF aims to promote a system to cost and prioritise climate actions; source and deliver finance; and track and make accountable expenditures (Finance Division 2014).

Once the government accepted the Sixth Five-Year Plan, the country's Outline Perspective Plan (Planning Commission 2010) was launched to achieve the national long-term goal for development called 'Vision 2021.' It projects a development scenario in which citizens will have a higher standard of living, better education, better social justice and a more equitable socio-economic environment. The sustainability of this development is to be ensured through a range of measures, including better protection from climate change and natural disasters (Planning Commission 2010).

In contrast to the Sixth Plan, the Vision 2021 document has approached migration in a more detailed manner. Like the Sixth Pan, this document welcomed remittances from overseas migrants. It raised concern about the large-scale urban migration that leads to a situation in which slums account for more than a third of the population in major cities. It cautioned against unplanned urbanisation and migration undermining rural development and making cities more vulnerable. It favoured a reversal of urban migration by promoting better living conditions and job opportunities in villages and controlling migration. It advocated a policy to weaken the forces of pull and push and inhibit rural-to-urban migration. The document shared a notion prevalent in development circles that migration from villages shows a failure in rural development.

However, later documents of the Planning Commission showed a change in the government's attitude to migration. In its 2012 progress report on the Millennium Development Goals (MDG), for instance, GoB favoured mainstreaming migration into development, climate change and environment policy. This argument was framed as part of the government's sustainable development discourse. This was also part of the government's efforts to streamline plans to offset the impacts of environmental change and human mobility challenges over the coming years (Planning Commission 2012).

The report noted that the population density in slums is many times the average population density of Bangladesh, which is among the highest in the world. "Bangladesh faces a Herculean task in sustainably improving the lives of slum dwellers" (Planning Commission 2012: 81). The report called for better lives for the slum dwellers. The report talked about assisting legal migration, facilitating remittances and access to labour markets for international migrants. International migration remained a focus of attention for the government. The report underscored the importance of remittances from international migrants as a driver for development and special measures to secure the rights of women international migrants. The report framed urban migration as a challenge for the cities, putting pressure on their services and infrastructure that are already stretched to limits.

In most of the development-related policy documents that we analysed, the narrative of internal migration is portrayed as a problem that needs a solution largely from a welfare point of view, or putting in place control measures and development measures that limit large-scale movements to cities. At the implementation level, the government tested resettlement projects for people affected by climate- and environment-related hazards and disasters. One classic example is the Guchhogram Climate Victims Rehabilitation Project. This three-year project was launched in 2009, aimed "to settle the climate victims, landless, homeless, address-less and river eroded people" (Guchhogram 2010). The project envisaged rehabilitation of 10,650 landless families in 207 villages in revenue land or land donated by a third party. As a result, 8,958 families have been rehabilitated in 198 villages spread across the country by 2013 (Guchhogram [CVRP] Project 2013a).

Set against the sheer scale of climate- and environment-related displacement, critics often dub the Guchhogram project as too small (YPSA 2013). However, it serves as a model for planned resettlement with a certain emotional appeal for Bangladesh's politicians and people. It relates directly to the liberation struggle of Bangladesh back in 1970, when the erstwhile Pakistan provided inadequate relief amidst political discrimination against its eastern wing, then called East Bengal (Heitzman and Worden 1989). After a violent civil war, and the country's liberation in 1971, Sheikh Mujibur Rahman became the country's first Prime Minister. He was keen to rehabilitate people affected by the cyclone. In 1972, the Ministry of Land Administration and Land Reforms rehabilitated 1,470 landless families affected by the cyclone and riverbank erosion in the *char* areas of the Greater Noakhali district. Later the government rehabilitated 45,647 families in 1,080 *adarsha grams* (ideal villages) in Adarsha Gram Project-I, and 25,385 families in 427 villages in the second phase of the project (Guchhogram [CVRP] Project 2013b). The appeal of the resettlement project continued, and the model was used in several intiatives for landless families. Though land and housing promoted earning opportunities, acces and right over land often came in the way of conomic pgorammes for the relocated families (Sultana and Mallick 2017).

Internationally planned resettlement of people affected by climatic and environmental hazards is an option explored in all seriousness in academic and policy circles (de Sherbinin et al. 2011; Ferris 2012, for instance). Migration from hazard-prone and underdeveloped areas of Bangladesh, especially its coastal zones, appears to be a clear case that warrants attention in this regard, especially in the context of people ending up in urban slums (MOEF 2008; Kreibich 2012; Kang 2012; Marshall and Rahman 2013; Muzzini and Aparicio 2013). At the same time, it may be noted that such movement is not always displacement, but more often than not, it is migration in search of better income.

6.4 Policies that address disaster vulnerabilities

A shared narrative in the development-related policy document is the hazard and disaster vulnerability of many parts of Bangladesh, especially its coastal areas, and its role in impoverishing people. One of the policies that squarely addressed this disaster-development nexus is the Coastal Zone Policy (MoWR 2005). This acknowledges the climate- and environment-related vulnerability of the coast, and cautions that disasters degrade the environment and impoverish people. It aims at poverty reduction through conservation of natural resources, promotion of sustainable livelihoods and provision for renewable energy on the coastal zone. The policy takes an integrated process approach and favours participatory decision-making and private sector participation to promote the development of the coastal zone. It recommended robust interventions to protect people from hazard exposure in line with the disaster risk reduction measures in vogue. The solutions recommended include early warning measures, protection against coastal and riverbank erosion and construction of dykes. It recommends rehabilitation of people affected, but still keeps silent about migration.

While the Coastal Zone Policy addresses the special vulnerabilities of coastal areas, the hazard-prone nature of the country itself, as described in chapter 2, has received policy attention in Bangladesh. In collaboration with UNDP, the Government initiated a transition from response and relief to comprehensive risk reduction in 2000. The outcome was the Comprehensive Disaster Management Programme (CDMP) that was approved in November 2003. CDMP was aimed primarily at disaster risk reduction; however, it incorporated climate change adaptation as a key component. CDMP Phase I (2004–2009) was a pilot for long-term disaster risk reduction and climate change adaptation in seven districts.

The initiative evolved after a decade of work following the devastating flood of 1988 and the cyclone in 1991. There were government and NGO initiatives that addressed disaster risk reduction in Bangladesh over the 1990s, and casualties in disaster became considerably lesser in later floods and cyclones. The government wanted to consolidate these gains and make an institutional framework for continued work in this regard. Based on this experience, the government preferred long-term planning and strategy to address disasters in a comprehensive manner.

CDMP I (2004–2009) became a pioneering initiative that laid the foundations for risk reduction approaches and frameworks in Bangladesh. It preceded the Hyogo Framework for Action 2005–2015 that was formulated at the 2005 World Conference on Disaster Reduction. The Hyogo Framework for Action 2005–2015: Building the Resilience of Nations and Communities to Disasters (HFA) was a pioneering mechanism to understand and promote multi-sector efforts to reduce disaster losses. Aiming at substantial reduction of disaster losses by 2015, it promoted a common programme to build the resilience of nations and communities. However, at the national level, CDMP I activities eventually related to the Hyogo Framework promoted further policy and planning mechanisms. It also facilitated the use of funds under climate financing mechanisms. CDMP took into account the underlying causes of vulnerability, hazard exposure and local contexts to promote effective disaster preparedness, prevention, mitigation, response and recovery measures (UNDP Bangladesh 2009).

The second phase, CDMP II (2010–2014), was envisaged as a vertical and horizontal expansion of CDMP Phase I. It addressed Bangladesh's climate and hazard vulnerability especially in the context of climate change, mainstreaming disaster risk management with climate change adaptation. Coordinated by Ministry of Disaster Management and Relief, the programme covered 13 key ministries and channelled funding through government and development partners, civil society and NGOs. It uses the PFM system and connects with the community and at the general stakeholder level through local governance structures (MDMR 2015). CDMP-II was supported by the United Nations Development Programme (UNDP), DfID, European Union, Norwegian Embassy, Swedish International Development Agency (SIDA) and Australian Agency for International Development (AusAID). The second phase of CDMP that started in 2010 with US $70 million includes a wide range of activities that address climate change impacts, as project areas are located in districts that are vulnerable to natural hazards. Phase II (2010–15) activities included mapping of hazards, risks and vulnerabilities and setting up an early warning system (MOFDM 2008, MODMR).

One of the outputs of CDMP II was effective management of community-level adaptations to disaster risks from a changing climate. The activity for this was collaboration and coordination with the SAARC (South Asian Association for Regional Cooperation) Disaster Management Centre and other donor countries to develop a national strategic plan to "address the challenges of climate change migration, refugees and displaced persons" (UNDP Bangladesh 2009). The outcome was the National Strategy on the Management of Disaster and Climate Induced Displacement.

The strategy is part of the Social Development Framework (SDF) of Bangladesh. SDF is a mechanism that covers the country's poverty reduction strategies on several fronts – education, health, nutrition, population, sanitation and water supply, financial inclusion, women and gender empowerment, social inclusion of ethnic and religious minorities, environmental protection, climate change management, disaster management, social security and overall sustainable development. Within this framework, the strategy paper takes a rights-based approached to climate-related migration and displacement – recognising the migrants' rights to safety, life and development (Siddiqui et al. 2015).

The strategy paper proposes a displacement management framework to facilitate responses during different phases of displacement. The paper mandates the state to protect the vulnerable people through better climate change adaptation and disaster risk reduction and management (Siddiqui et al. 2015). It favours preventing displacement, while ensuring that migration and displacement that are inevitable are managed. During the pre-displacement phase, it recommends understanding of risks, investments in disaster risk reduction and climate change adaptation, strengthening of disaster risk governance, and promotion of livelihoods and income generation, including through decentralised urban growth and climate-disaster risk responsive land use plan. During the displacement phase, the strategy calls for quick and decisive intervention to manage it, address urgent humanitarian needs and ensure effective protection. The post-displacement phase intervention seeks to avoid protracted situations through durable solutions – return, local integration and relocation/resettlement (Siddiqui et al. 2015).

The strategy paper takes into account the state of the art in climate-related migration and displacement research and policy-making. In the context of Bangladesh, with its large-scale urban migration, it will be an ambitious task to implement this strategy. However, the paper itself calls for effective data gathering and raising funds through climate financing and through development channels. The paper has effectively filled the glaring policy gaps in addressing migration in other documents that deal with development and disasters in the country.

As for dealing with disasters as such, the key policy documents include the National Plan for Disaster Management, the Disaster Management Act and the draft National Disaster Management Policy. In particular, the National Plan for Disaster Management 2010–2015 is the blueprint for action in this field. In turn, operational aspects have been codified in the Bangladesh Disaster Management Act that was approved by the Minister for Food and Disaster Management in 2012 (MDMR 2013). It was first drafted in 2008, and revised and enacted in 2012. Under the act, the Department of Disaster Management (DDM) was set up under the Ministry of Disaster Management and Relief to reduce the overall disaster vulnerability by coordinating risk reduction, emergency response as well as humanitarian assistance (DDM 2012). The policy-making process has been undertaken with NGO and community participation in the second phase (2010 to 2014) of the country's Comprehensive Disaster Management Programme (CDMP). The CDMP focuses on risk management and mainstreaming.

A comprehensive document that covers Bangladesh's disaster response measures is the National Plan for Disaster Management 2010–2015. It addresses disaster risks comprehensively, suggesting ways to reduce the vulnerability of poor people to the effects of natural, environmental and human-induced hazards. It aims at "(1) bringing a paradigm shift in disaster management from conventional response and relief practice to a more comprehensive risk reduction culture and (2) strengthening the capacity of the Bangladesh disaster management system" (PreventionWeb 2012). The plan document (DMB2010) relates to climate change and development policies of the country discussed above. It also explains how different parts of the country become vulnerable to disasters. It analyses the socio-economic dimensions of disasters and acknowledges the poverty-disaster interface and the impact of disasters on economic and social activities of the poor. It lists depletion of assets, reduced income due to loss of work and increased indebtedness as factors that increase the vulnerability of the poor. It also notes how the cost to cope is disproportionately higher for the poor (p 36), and acknowledges that floods and riverbank erosion are rendering people homeless. Still it falls short of addressing the option of migration as a strategy for poor people to cope with disasters or to adapt to climate change. It includes outmigration as a factor that increases the vulnerability of the poor. Resettlement does receive a passing mention: "Disseminate the information for utilization in development planning and resettlement of vulnerable communities" (p 73). Yet resettlement – usually a top-down state-led process – is not the same as migration as an adaptation strategy for poor people.

Similar to the National Plan, the Disaster Management Act and associated policies also address disaster risk in Bangladesh and make positive contributions to reducing loss of lives and livelihoods in account of disasters, but they largely overlook migration.

6.5 Migration policies

While the policies discussed above looked at how climate change, development and disasters influence migration, there is a set of migration policies in Bangladesh that deals largely with labour migration. The focus here is to examine to what extent they address climate- and environment-related migration. The government policy on migration builds on the 1982 Emigration Ordinance that was devised at the height of labour demand in the Gulf States. As labour dynamics changed, there have been periods of restrictions on migration, notably in response to the exploitation of migrant workers, and especially migrant women. However, over recent years, policy has become much more open and facilitating, partly due to demand from migrants themselves and the recognition of the role that migration plays in the country's economy.

One example of this is the Overseas Employment Policy, 2006, which sets out the right of male and female workers from Bangladesh to choose overseas employment. It is aimed at regularising migration from all parts of the country. It protects the rights, dignity and security of the migrant workers within and outside of the country. It also seeks to ensure social protection of their families that stay back and commits to strengthen the policy implementation mechanism (Siddiqui and Farah 2011). A more recent legislation, titled the Overseas Employment and

Migration Act 2013, upholds and protects migrants' rights, based on the principle of non-discrimination. It allows supported emergency return of migrants if their host country is in any kind of crisis. It tries to prevent fraudulent practices and enforces accountability of recruiting agencies and their sub-agents. It also has a provision for functioning sub-agents. In the past migrants could not go directly to court against the misconduct of a recruiting agency. The new law allows a migrant to move to court if the concerned government official fails to take legal action in time in such cases. Before its enactment, the draft law has gone through civil society consultation, and was presented to the Ministry of Law, Justice and Parliamentary Affairs (Siddiqui and Farah 2011; Siddiqui 2011). Apart from the above provisions, this law enhances the safety of women's migration, and makes recruiting agencies more accountable (MOF 2014).

6.6 Climate change policies

After drawing up a national framework as mandated by the UN Framework Convention on Climate Change (UNFCCC) for the least developed countries (LDCs), Bangladesh submitted its National Adaptation Programme of Action (NAPA) in 2005. The document noted that climate change will increase the impacts of natural hazards in the country, and made an urgent call to integrate adaptive measures into the development process (MOEF 2005). Based on the broad directives of NAPA, the Climate Change Unit of the Ministry of Environment and Forests (MoEF) brought out the Bangladesh Climate Change Strategy and Action Plan (BCCSAP) in 2008. Both the documents were revised in 2009 after a government change.

To follow up on projects envisaged by these documents, the government set up a Climate Change Trust Fund (CCTF) in 2009. The CCTF has approved 43 government projects worth US $70 million, besides 32 NGO projects worth US $3.5 million. The government has also put in place the Bangladesh Climate Change Resilience Fund (BCCRF) – based on public finance – with development partners pledging US $113.5 million. The resilience fund is being managed and implemented by the Government with the World Bank's technical assistance. The government has set up a multi-donor Trust Fund (MDTF) to manage adaptation funds (UNFCCC 2014). During 2009–12, the government allocated US $350 million from its non-development budget and approved 107 projects worth an estimated 1,272 million Bangladeshi Taka (US $16.4 million) in areas that are vulnerable to climatic stresses and shocks (Pervin 2013). The government also draws from multilateral development banks and agencies for low-carbon resilient development (LCRD) investments. These donors include the World Bank, ADB and International Finance Corporation. Other major intermediaries include the Global Environment Facility (GEF) and national banks, including the Central Bank (Pervin and Moin 2014).

Being the first key document mandating action in the face of climate change, the NAPA in its original form (MOEF 2005) set the tone for adaptation activities in Bangladesh. The NAPA narrative on migration, however, is characterised by an approach in which migration is assumed to be problematic. It treats migration as an undesirable outcome of climate change. For example, a diagram in the document mentions migration along with crime as an outcome of livelihood impacts of climate change (MOEF 2005: 17). In the list of projects, one promotes ways to adapt coastal farming to increasing salinisation. It envisages adaptation to floods, storm surges and sea level rise. Moreover, it adds as project outcomes: "Affected community would not migrate to cities for job and livelihood" (p 35); and "Social consequences of mass scale migration to cities would to some extent be halted" (p 36). Another project talks about the need to undertake adaptive measures in the north-east and central regions that are often flooded. As the document notes: "In the long-term, people might get a means to continue with farming instead of migrating to cities after the flood. This would to some extent reduce social problems of migration of the distressed community to cities" (p 37).

The Awami League government that came to power in 2008 supported the NAPA and BCCSAP prepared by the preceding government, but made its own contributions in their newer version, as our key informant interviews in Bangladesh suggest. They selected migration as the issue to show this distinctiveness: most of the negative references to migration in the original NAPA were deleted, with the exception of one diagram on p 17. However, although the NAPA and BCCSAP were updated, still the revised documents did not see migration as an adaptation strategy worth promoting. The updated NAPA focused on four security issues, namely, food, energy, water and livelihoods, and respect for local community in matters related to resource management and extraction (MOEF 2009). The list of projects related to climate change was expanded further.

While the NAPA mainly lists immediate priorities, the BCCSAP focuses on medium and long-term goals under six broad areas. These are food security; social protection and health; comprehensive disaster management; infrastructure research and knowledge management; mitigation; low carbon development; and capacity building and institutional strengthening. In line with the original NAPA, the original first version of BCCSAP (MOEF 2008) warns of grim scenarios of climate change. For instance, it warns: "unless existing coastal polders are strengthened and new ones built, sea level rise could result in the displacement of millions of people" (MOEF 2008: 1). It is due to farms becoming less productive with restricted livelihood options. Such migration, especially from the coastal zones, is estimated to comprise 6–8 million people by 2050 (MOEF 2008). City slums, their likely destinations, are a potential problem given the fast, but unplanned urbanisation of the country (ibid: 16). Still BCCSAP did not outline any policy response to such migration.

In its 2009 update, the BCCSAP has addressed migration in more detail, updating the figure to a staggering 20 million, suggesting that these 'environmental refugees' have to be resettled, possibly abroad. "Migration must be considered as a valid option of the country," the updated version notes. "Preparations in the meantime will be made to convert this population into trained and useful citizens for any country" (MOEF 2009: 17). The section on Research and Knowledge Management suggests "monitoring" of climate change-related internal and external migration and rehabilitation (MOEF 2009: 58). The key climate policy documents start with a pessimistic view of migration, but later their editions soften this stance by acknowledging migration as a viable option. While the original NAPA and BCCSAP documents seek policy interventions to either reduce the need to migrate or deal with it as a form of forced displacement when it happens, the updated 2009 version explores options to facilitate migration by training them and suggesting resettlement options.

6.7 Discussion

Growing urbanisation and industrialisation in Bangladesh includes migration as part of a 'single symbiotic process' (Marshall and Rahman 2013: 5). If that is the case, internal migration – along with international migration – is necessary for economic growth and poverty alleviation in the country. The Government of Bangladesh has taken this approach in its recent policy documents. The MDG Progress Report 2011 (Planning Commission 2012), for instance, places migration in the context of economic development, as well as environmental concerns. Meanwhile, the Sixth Five-Year Plan (Planning Commission 2011), Ten-Year Perspective Plan (Planning Commission 2010) and the National Strategy for Accelerated Poverty Reduction (Planning Commission 2009) acknowledge the importance of short-term international labour migration in the economic development of the country.

However, internal rural-to-urban flows, especially to the metropolitan core, are not generally encouraged so much in government documents. There is a growing recognition of internal migration in some of the policy documents (Planning Commission 2011, for instance). The stress is more on forced migration, and the responses such as the Guchhogram project are largely reactive. In addition, whilst the overall view of migration in Bangladesh policy circles is becoming more positive in terms of its potential to promote development, a glaring omission in most of the policies discussed above is a lack of mention of internal migration or its portrayal in negative or, at best, general terms. This has a significant impact on how migration is seen as a phenomenon related to climate change, as policymakers are now trying to bundle climate change concerns – from disasters to gradual deterioration of the environment and loss of livelihoods – and address them in comprehensive policies. One issue is an apparent conflict of interests between the poor people who migrate to cities in search of work in large numbers and industry and business. The use and ownership of urban space is a contentious issue. The Sixth Five-Year Plan document, for instance, talks about a tremendous pressure on land and natural resources exerted by the migrants (Planning Commission 2011: 188–189). The Outline Perspective Plan seeks to "reverse" the trend of migration (p 68) and put in place "migration controls" (p 69). In the plan document, there is no mention of the role of migration in delivering workers to the growing industrial sector, never mind its role as an adaptation to climate-related shocks in rural areas.

Indeed, although a city-based development pattern in Bangladesh draws people from villages in large numbers, surprisingly absent from government discourse is a proper acknowledgement of the economic contribution of internal migrants and their role in a growing economy. There is mention about changing farming practices and growth of cities – but this knowledge is not translated into an enabling environment for the migrant workers. There are exceptions such as the Guchhogram Project. However, the contribution of migrants to the economy largely goes unnoticed. Research and policy measures often ignore unorganised temporary migrant workers, their agency and rights (Rogaly 2009).

At the same time, there is widespread awareness and a strong political will in policy circles in Bangladesh to acknowledge the impacts of climate change on development and growth. The Sixth Plan document notes: "climate change will exacerbate the vulnerability of poor people to environmental shocks, with the predicted increase in extreme climate events" (Planning Commission 2011: 165). This understanding is part of an effort of the government's effort to mainstream concern regarding climate change into the overall planning process. Climate change is not being treated as merely an environmental issue, but a development issue. In turn, the country has developed a more proactive stance towards migration that sees this also as integral to development, even if the focus to date has been on international rather than internal migration.

Even though many policies in Bangladesh are progressive and people-friendly, there is some concern about a lack of co-ordination among different ministries with policies tending to take a silo approach, not accounting for issues addressed in different, but related policy areas. For example, the Overseas Employment Policy that is pursued by the Ministry of Expatriates' Welfare and Overseas Employment (MOEWOE) does not deal with climate change issues; whilst environmental policies pursued by Ministry of Environment do not look into the broader aspect of labour migration that the government is promoting. There is no monitoring and evaluation process built in to the policy (Siddiqui and Farah 2011), although the government is collaborating with international agencies to better understand climate- and environment-related migration (ADB 2011). IOM has also initiated policy dialogues on mainstreaming concrete short and long-term migration adaptation strategies in Bangladesh based on emerging evidence.

Another aspect of migration that does not find mention in policies is irregular cross-border migration. This is a thorny issue in India-Bangladesh bilateral relations, especially after India resorted to tough border control measures in recent years (AP 2012). The issue has cropped up in several bilateral meetings of security agencies, and at the Home Minister level. A Coordinated Border Management Plan between the two countries aims to curb illegal border crossings and incidents (Rajya Sabha 2012), even as international agencies such as the IOM are trying to use provisions for cooperation under the South Asian Association for Regional Cooperation (SAARC) Convention to promote a more flexible approach. Bangladeshi migrants are also reported to travel to Malaysia through the Bay of Bengal and then jungles of Thailand (Siddiqui 2014). In 2014, about 54,000 people migrated to Malaysia through this route, 53,000 of them Bangladeshis; 540 people are estimated to have died attempting the passage, mostly due to starvation, dehydration and violence by crew (UNHCR 2014). In the first quarter of 2015, about 25,000 people have departed from Bangladesh and Myanmar in irregular maritime movements from the Bay of Bengal, almost double the number from the same period the previous year; an estimated 300 people have died at sea (Tan 2015). These reports show the risky nature of international migration for many Bangladeshis that are often inadequately addressed in policies. Since cross-border and international migration also involve climatic and environmental influences, they need to be addressed as part of a comprehensive migration policy regime.

A close look at migration patterns in Bangladesh shows that more often than not, migratory movements are within the country. Often they involve short-term shifts to neighbouring places that are familiar to the migrants. Village-to-village migration and often urban forays help the migrants supplement their livelihoods and tide over tough phases and lean seasons. Often people move back and forth, over long and short distances. Mobility broadly means the freedom to seek opportunities to improve livelihoods, living standards and services such as health care and education – succinctly put, safer and more productive life in more responsive communities (UNDP 2010). It is a broader concept than migration, a fundamental element of human freedom (UNDP 2009).

Migration experts increasingly suggest that policies that acknowledge and promote human mobility succeed and those that restrict it fail (de Haas 2009). Yet despite some movement towards recognising this, our analysis of key public policies in Bangladesh suggests that the country remains some way from acknowledging the potential benefits of migration for adaption to climate change in practical ways. Efforts at such a policy realignment could start by acknowledging voluntary labour migration – both internal and international – as a way to improve the resilience of climate vulnerable communities and an effective adaptation strategy. Beyond the strategies and projects proposed in the NAPA and BCSAS, the government could also frame a comprehensive climate change policy aimed at climate-resilient development, which appreciates the role of migration as a climate change adaptation strategy and addresses the hardship of displaced people. To ensure that migrants benefit from such a policy, reforms are also needed across different socio-economic and development policies. There is a need to establish the habitat rights of the displaced people in *khas* (state-owned) and diluvian (flood-related) land. At the same time, labour policy should be made more comprehensive and inclusive to protect the rights of internal migrants – ensuring living wages, access to health care, social safety net and civic services, and safer working and living conditions. Such protection is especially important in construction, garment and brick kiln industries, where people from climate-vulnerable areas often find work.

Meanwhile there is a need to focus on areas affected by climatic hazards. A review of the Overseas Employment Policy 2006 would make it easier for people from such areas to obtain short-term international migration contracts. Vocational training facilities such as technical training centres of the Bureau of Manpower Employment and Training (BMET) and the ministries of youth and education could be extended to these areas. The government-run Prabashi Kallyan Bank (Migrants' Welfare Bank) might also offer better services in such areas to provide loans for prospective migrants and to promote enterprises using remittances, thereby enhancing job opportunities. Such creative use of remittances can contribute to local resilience and rebuilding after disasters (Stern 2007). To streamline such intervention, labour and overseas employment ministries would need to have representation in government committees dealing with climate change, such as the Inter-ministerial Climate Change Steering Committee, the technical committee of the Climate Change Trust Fund (CCTF), National Environment Committee, National Committee on Climate Change and Climate Change Unit.

Policy realignment is a political process that requires procedural and administrative interventions. Given the political will, the commitment of the national government to put in place effective climate change adaptation mechanisms (UNFCCC 2014) and the administrative facilitation of such measures as described above, a reform process that enhances the adaptation role of migration is certainly possible.

6.8 Conclusion

This book argues that villagers in Bangladesh migrate for better livelihoods, sometimes in response to climatic stresses and shocks. Most of this movement is within the country and if facilitated appropriately, such internal migration – along with international migration – can be an effective adaptation strategy. It can help build individuals' and communities' adaptive capacity to future environmental and climatic hazards, while migration happens in the context of a growing city-centred economy that promotes remittances to villages. However, a textual analysis of current and recent policies concerning climate change, development and poverty alleviation, and disaster management shows that the economic and adaptive roles of internal migration are often not included in policy documents.

Though recent documents acknowledge that there is a great deal of migration in and from Bangladesh, the responses outlined are largely reactive rather than proactive. Migration is often framed as a problem (for the migrants), outcome of a problem (at the migrants' place of origin) or cause of more problems (at the destinations). So policy measures tend to be prescriptive, often attempting to restrict or discourage migration by various means, including rural development and alternative livelihoods, or dealing with problems at the destination. At the same time, city-based growth encourages migration at a large-scale. In that case, migration is part of the economic activity that helps people offset the impacts or rural poverty as well as climatic stresses and shocks and environmental degradation. Here the argument is that if migration works as a positive step towards adaptation, then the key challenge is to realign the policies with this new understanding.

7 The song of the road

Weaving together the migration story

While placing these findings within the scholarship in the field of climate- and environment-related migration, this chapter narrates a storyline on how climatic and environmental stresses and shocks as well as perceptions of the risks they pose influence migration decisions of people in Bangladesh. It also explains how a multi-method study of migration decision-making in a climate change hotspot relates to international trends and the climate migration story in general. It makes the point that though poverty drives migration, the poorest are often unable to move out, in effect possibly getting trapped in environments of potential hazard risk or moving to even riskier places. It makes the case that irrational restrictions placed on people's mobility options could lead to trapped populations vulnerable to hazards. It calls for more proactive migration policies.

This chapter synthesises the findings of the thesis, draws new lessons and identifies some of the uncertainties that it leaves behind in its effort to understand climate- and environment-related migration in Bangladesh. It draws from the key findings discussed in the three empirical chapters (chapters 4–6) and relates them to the case study (chapter 2) of Bangladesh and the broader context discussed in the literature review (chapter 3). While placing these findings within the scholarship in the field of climate- and environment-related migration, it traces the storyline of people responding to climatic and environmental stresses and shocks as well as perceptions of the risks they pose – by staying or moving. The story set in the context of a country facing frequent and devastating climatic stresses and shocks, a climate change hotspot (Huq 2001; Huq and Ayers 2008), can be of possible academic and policy and relevance, and contribute to opening up future avenues of enquiry.

First this chapter pools together key findings from the empirical chapters – qualitative, quantitative and policy analyses – and weaves them into a single, coherent narrative. Second, it traces the linkages of different strands of this narrative to the literature and identifies how they challenge or chime with their findings. Third, it examines how migration in Bangladesh can be seen as a climate change adaptation measure, as well as how to ease restrictions to mobility – on account of people's abilities and choices, opportunities and resources they have, and the policy environment – that could in effect trap vulnerable people in potentially hazard-prone places. Summing up, it acknowledges the limitations of this research effort, uncertainties involved and explores ways to engage further in this field.

7.1 Synthesis of the empirical chapters

7.1.1 Socio-economic factors that drive migration

The empirical chapters together give a multi-dimensional view of climate- and environment-related migration in Bangladesh. The qualitative analysis (chapter 4) narrates mass movements of people from across the country, especially its less developed coastal belt, which is also exposed to climate- and environment-related hazards such as floods and cyclones (Marshall and Rahman 2013; Muzzini and Aparicio 2013). The respondents in general note that usually it is income needs that drive their migration in the context of rapid urban growth. People living in areas exposed to climatic stresses and shocks – floods, droughts, cyclones and riverbank erosion – often have to diversify their livelihoods to ensure better income and sometimes to offset losses suffered on account of these hazards. Farm-based livelihoods still constitute a major share of Bangladesh's economy, but it is shrinking.

The quantitative analysis supports this narrative of poverty-driven migration with numbers. Poverty comes across as a significant variable that positively influences migration – though it is not necessarily the most significant factor. While poor people leave an environment that does not support their livelihoods adequately, many draw on their assets for survival at their place, staying put, without migrating. It appears to be difficult for people without assets to undertake migrations outside their district of origin, as the quantitative analysis finds. Long-distance migration requires a significant amount of resources that people without any assets may not be able to gather easily. The study also shows that though poverty drives migration, the poorest are often unable move out, in effect possibly getting trapped in environments of potential hazard risk.

While it is poorer people in villages with assets who tend to migrate, they draw from the experience and support of friends, family and peers who have undertaken similar journeys. The qualitative analysis shows that often social networks play a very important role in migration, by telling people about opportunities and the way to go about migrating, financing and facilitating their journeys. People often follow the migration routes charted and undertaken by their friends and family, neighbours and colleagues. The migrants trust the accounts of their social network members more than any other source – governmental or otherwise – while making migration decisions. Such accounts often influence the direction of migration and the choice of livelihoods at their destinations, as the literature shows. This narrative is strongly supported by all the models in the quantitative analysis. Social networks come across as a highly significant variable that positively influences migration outside the district as well as moving house.

It is clearly younger people, who have better chances of employment, and those with more education who tend to migrate more. However, the role of education was not very clear in the qualitative study, though that of age was clear. It is often younger respondents who talked more emphatically about leaving home for work outside. Education, however, has a negative impact on migration within the district, as the quantitative analysis shows. This could be because migration outside the district is to centres of economic activity, clearly a choice for better economic opportunity. Possibly, respondents who are more educated are better placed to undertake such endeavours than others, as the literature suggests. Movements within the district possibly denote more of the same livelihood activity under better circumstances, so education does not really add value to such a shift.

Even while migration decisions are made mainly for better livelihoods and income, in the background of climatic and environmental stresses and shocks, they are not solely based on consideration of cost and benefit (Massey et al. 1993) or risk and resilience (Wisner et al. 2004). The processes are far more interlinked and complicated than any simplistic, monocausal model can predict. People experience changes around them, perceive the risks they pose and assess the response options before them, as the qualitative analysis shows. The respondents' own attitudes as well as socio-economic backgrounds, cultural practices and social norms influence the choice of options before them, as the qualitative analysis shows. Migration decision-making is a deliberate act that involves agency, but is mediated by a set of social and cultural variables.

The qualitative analysis frames migration as a socially acceptable behaviour that people engage in for economic gain in the context of their experiences of the local climate and the environment and the perceptions of risk they pose. These decisions are also influenced by a set of behavioural components. In this framing, migration across different spans of time and space is a strategy to diversify livelihoods for better income. So households often diversify livelihoods by sending one or some of the members away to work and thus reduce their vulnerability to shocks and stresses, including climatic ones. At the same time, the prevalent local narrative of migration is all about improving income. A hostile environment, even when environmental stresses and shocks make livelihoods increasingly insecure and unsafe, works as a grim background of the migration in which the migrants are heroes, not victims. In short, in a range of time-space combinations, migration contributes to local resilience and climate change adaptation, though migrants themselves do not use these terms or explicitly acknowledge these notions. This aspect is discussed in detail in sub-section 7.2.3.

Migration decisions are taken firmly and deliberately, the respondents carefully weighing the pros and cons of migration against other options, as the qualitative analysis finds out. This finding follows up on earlier studies that reached similar conclusions (Kniveton et al. 2008; Smith et al. 2010; Kniveton et al. 2011). Even when there are climate- and environment-related threats, the decision-making process involves testing the possible options and outcomes – through comparison with experiences of family, friends, neighbours and peers. This may not be the case in the case of displacement after sudden events such as riverbank erosion or a cyclone. The creative and bold actions involved in migration – with no external support – suggest that people have a sense of control over their destinies. Even while there is a shared belief in the pre-destined nature of disasters as acts of God, it does not hinder preparatory and remedial action.

7.1.2 Climatic and environmental factors that drive migration

The respondents, especially in the cyclone-affected Satkhira district and the drought-prone Nawabganj, talked about moving out of their place after heavy losses suffered because of climatic shocks and stresses. It is not only a one-off event of a severe drought or a devastating cyclone that pushes out people, but also their lingering effects, such as dipping groundwater levels, salinity intrusion after a storm surge and fallow farms that lead to loss of farming and fishing livelihoods and labour opportunities. In Munshiganj, prone to floods and erosion, the story is often that of sudden displacement after the local river swallows up part of the farm, homestead or the house itself.

Looking at the tales shared by respondents in the qualitative part of the study more closely, it appears that while it is income needs that drive migration, landless and marginal farmers tend to migrate more than those with better means, especially in the aftermath of weather uncertainties and associated losses. Combined with findings from the quantitative analysis, it appears that such movements often involve short distances, within the district, or short-term or seasonal trips outside the district. The long-term moves outside the district are driven by poverty, but supported by assets. That means migration often becomes a fallback mechanism for poor people or coping strategy at times, especially after disasters. Still, it becomes difficult for the poorest to migrate to a faraway city to earn a better living. Poverty and income needs are cited as the main reason behind migration – an observation validated in the quantitative analysis, as explained below. Still climatic and environmental factors often impoverish people, or make people poorer, as the qualitative analysis notes.

As villagers in Gabura recalled, their farms were inundated in the storm surge after the 2009 Cyclone Aila. Three years hence, at the time of the fieldwork for this book, Aila's legacy still lingered on in Satkhira district. The villagers could not grow rice in their farms due to salinity left by the storm surge that had inundated the village. Often the solution for problems such as climatic and environmental stresses and shocks, as the focus groups reveal, is seasonal migration, with men working in faraway farms or in towns. In Gabura, the local women said that the men go to work for three to four months at a stretch to work in farms within or outside the district or in search of opportunities for casual labour in cities. Meanwhile their village still remained vulnerable to weather extremes, exposed to fierce storms and possible inundation if the storm surge breaks its mud embankments.

In such a context of hazards and their lingering effects, migration becomes a strategy to deal with multiple stresses. The quantitative analysis throws more insights into this multi-causal nature of migration. It confirms that migration is largely a poverty-driven phenomenon, also influenced by stresses and shocks such as cyclones, floods, rainfall variability and droughts. Cyclones and droughts appear to push inter-district migration, reflecting the interview narratives of the respondents of Satkhira and Nawabganj, respectively, in the qualitative part of this study.

Migration of families and women from the coastal areas has been explained by the overall disaster vulnerability of the coastal region in recent literature. There has been large-scale migration from the coastal areas affected by Cyclone Aila, and often whole families migrated due to livelihood stresses (Poncelet et al. 2010; Islam and Hasan 2016). Recent literature shows that such migration is different from displacement in the immediate aftermath of destruction or inundation of habitat, but driven by livelihood needs. People tried to live in their places of origin, at least a week after the disaster; however, scarcity of water and food crisis led to long-term migration to urban areas, especially Khulna city and its peripheries (Islam and Hasan 2016). A factor that adds to migration is increased stresses in farming livelihoods. For instance, recent modelling identifies salinity and temperature stress reducing crop productivity, and private debts further impoverishing farmers in the coastal area (Lazar et al. 2015). Studies in coastal Bangladesh show that women's livelihood activities, including gathering natural resources and protecting family assets (including farms, ponds, poultry and cattle) can be severely affected by climate variability and change (Garai 2016).

Meanwhile riverbank erosion prompts short-distance, largely local-level movements. In the north-western region, at Chorpka village of Nawabganj that is prone to riverbank erosion, a 35-year-old woman narrated her experience of forced migration: "In 1998 we were displaced due to flood and riverbank erosion. At first we migrated from Radhakantapur to Sohimullah village. Again we faced the same disasters in 2000, 2004 and 2008 there, so we came to Chorpka village." Such multiple displacements are commonly reported from the district's riverine islands, as the literature shows. The quantitative analysis places riverbank erosion as a significant reason for people leaving their village, but not necessarily the district. The experience of erosion has a positive influence on moving house, but negative influence on inter-district migration, as the respective logit models show.

In poverty-related migration supported by social networks, climate- and environment-related experiences often play a strong positive or negative role, as the qualitative analysis shows. Respondents who have experiences of climatic and environmental stresses and shocks also tend to be concerned about the risks they pose, as the qualitative analysis shows. These experiences and concerns often necessitate diversification of livelihoods to enhance income. This risk concern factor could not be ascertained in the quantitative models possibly due to inherent limits of the tools used – in correctly assessing the level of concern about climatic and environmental events and processes.

In the qualitative analysis people said that such uncertainties, stresses and shocks affected their livelihoods, but seldom linked these experiences to their migratory movements. It is only in the case of sudden changes – such as loss of land due to erosion or flooding due to a cyclone – that people acknowledged that migration could be a coping strategy. Significantly, respondents almost never associated floods, however frequent and severe they may be, with long-term migration. Floods appear to discourage local movements as well as inter-district migration, as the quantitative analysis shows. The quantitative analysis shows that experiences of cyclones and droughts have largely positive influence on inter-district migration, but negative influence on local moves. On the contrary, riverbank erosion has a negative influence on inter-district migration and positive influence on local moves. This observation chimes with what has been shown in the qualitative analysis and in the literature discussed in chapters 2 and 3. While exposure to cyclones and droughts tends to drive migration for better or alternative livelihood pursuits elsewhere, riverbank erosion often involves displacement – often over a short distance, sometimes repeatedly.

A combined reading of the first two empirical chapters shows that it is not necessarily the hazards prevalent locally that drive migration but a variety of factors – largely poverty, scarcities and uncertainties – in combination. When people migrate for better livelihoods, sometimes in response to climatic stresses and shocks, such migration can be interpreted as an effective adaptation strategy. This book assumes that such migration can help build individuals' and communities' adaptive capacity to future environmental and climatic hazards. At the same time, migration happens in the context of a growing city-centred economy that promotes remittances to villages. The policy analysis (chapter 6) probes how the government views these movements, their underlying causes, consequences and implications; and attempts to facilitate, ignore or control migration. As such, the policy is largely silent, if not outright negative at times, about acknowledging the role of migration as an effective climate change adaptation strategy, as this book argues. Still, the country's economic policies promote city-based growth at a fast pace, thereby tacitly promoting large-scale rural-urban migration.

7.2 Linking with the literature

7.2.1 Different dimensions of climate- and environment-related migration

This book relates to the literature on climate- and environment-related migration in three dimensions. First, it underscores the multi-causality of migration as conceptualised in some of the neoclassical migration theories and recent influential empirical and synthesis studies. Even when climatic and environmental stresses or shocks influence movement decisions, as this book argues, it is primarily income needs that drive migration.

Second, the book illuminates the differential impacts of climate-related hazards in driving or inhibiting migration – across types of hazards and socio-economic categories of people. Besides, the effects of different disasters and hazards have a broad range influence on migration – from causing sudden displacement and prompting migration for better income to causing immobility. This finding is a clear departure from early literature about what has often been described as 'climate-induced' migration, and the notion of direct and positive correlation between climatic stresses and shocks and migration.

Third, as a follow up, the book stresses the implications of immobility of people in hazard-prone places, an issue that has gained attention in recent literature on climate- and environment-related migration. Closely connected with this issue is the way people perceive the influence of climatic and environmental hazards in their migration decisions. As the book blends people's experiences of climatic and environmental hazards with instrument observations to test the climatic and environmental sensitivity of migration decisions, it looks at these linkages from different vantage points. It finds that people usually do not associate their movement to the hazards that appear to indirectly affect their livelihoods, even when they influence movement patterns – or inhibit migration altogether.

Migration here is framed as a beneficial activity that helps people offset losses suffered as a result of climate- and environment-related stresses and shocks and be better prepared for such experiences in the future; as such, it can be considered an adaptive activity. On the basis of such a conceptualisation, the book considers the policy implications of its findings, especially on how restricting migration from hazard-prone places might make such immobility a big problem as vulnerable people might get trapped in such places.

7.2.2 Multi-causality of migration

The first point about multi-causality of migration has found expression in reviews and synthesis studies. Migration has been conceptualised as a varied and complex outcome of economic, social, environmental, demographic and political processes operating at different levels (Foresight 2011). In a matrix of poverty and inequity, as the environment changes and resources get used up, climate along with other socio-economic, environmental, political and demographic factors, could drive migration (Lonergan 1998). Foresight (2011) stresses this multi-causal nature of migration and outlines how migration patterns change as a result of the influence of global environmental change on multiple drivers of migration. Kniveton et al. (2008), for instance, explain migration as one among many ways that individuals, households and communities escape poverty. Theoretically, this view reflects what early migration theorists have proposed about the multi-causal nature of migration. As the literature review shows, migration has been conceptualised as an adaptation to perceived changes in environment, a way to escape from the marginal status in the migrant's place of origin in search of better opportunities in a chosen destination (Wolpert 1965).

Recent empirical studies in Bangladesh on climatic and environmental stresses and shocks and their impact on migration have focused on how income needs, rather than the hazard per se, drive migration. Gray and Mueller (2012), for instance, have shown that household-level economic damages averaged a fifth of household expenditures for flooding and 12 per cent for crop failure, though the recovery rate for households that have experienced migration crop failure was slower than that of flood-affected households. In effect crop losses, even when not associated with floods, led to more migration than when people faced flood alone. On another plane, Gray and Mueller (2012) stress the multi-causality of migration and argue that it is not the poor who always migrate, possibly due to multiple barriers. The inability of people without any assets to undertake long-distance migration is a recurrent theme in this book. The issue of immobility of the poor is a key concern raised in the book, as discussed below.

Related to the above analysis, but in what appears to be a contradictory note, Penning Rowsell et al. (2013) argue that population movements are driven by the need for safety and income recovery after hazard exposure, especially for landless people. In their study on cyclones in Bangladesh, Paul and Rautray (2010) note that migration is more prevalent among people in the lower income category than in higher and middle income groups. While this thesis shares the notion that among the multiple causes of migration income takes a predominant role, it further argues that people without assets find it difficult to migrate out of the district even though lack of assets is not a barrier for short-distance shifts.

7.2.3 Differential influence of climatic and environmental factors on migration

The second point about differential impacts of hazards in migration patterns finds expression in the logit models. They show that experiences of drought and cyclone appear to positively influence migration outside the district, even while these hazards have a negative influence on the first house shift that is largely within the district. Short-distance movement that happens because of these hazards – as reported in qualitative analysis and the literature – is likely to be short-term moves that do not involve shifting of residence. While riverbank erosion as well as floods negatively influence long-distance movement, people affected by erosion tend to move locally. Each of these observations finds reflection in the existing literature.

Recent literature shows strong evidence for seasonal and temporary migration in Bangladesh. Droughts, especially in northern Bangladesh, often lead to lean periods between harvests called *monga*, marked by poverty and food insecurity (Findlay and Geddes 2011). Landless labourers, mostly boys and men from *monga*-affected areas, migrate seasonally to cities and better-off villages in search of work (Siddiqui 2009). Penning Rowsell et al. (2013) report that drought in the dry season translates into reduction in farm work opportunities, driving seasonal migration by men to other districts where there is a better demand for labour; however, permanent movement over the last decade involved only less than a tenth of the households. Disasters, especially, droughts, prompt temporary, short-distance moves in Bangladesh, but their influence on permanent migration is found to be minimal (Bohra-Mishra et al. 2014). Research elsewhere also associates drought with migration (Munshi K 2003 and Henry et al. 2004, for instance). In this book, drought-driven migration is a recurrent theme in focus groups and interviews; and drought comes across as a strong driver of migration outside district, as the quantitative analysis shows. While the qualitative analysis captures the extent of the temporary and seasonal movements because of drought, the logit models show strong evidence for long-term migration outside the district.

One hazard that appears to hinder long-distance and short-distance movement is flood, a yearly occurrence in many parts of Bangladesh. The literature shows that though flooding is frequent, widespread and damaging, it has only modest effects on mobility (Gray and Mueller 2012). A reason behind the subdued effect of flooding on long-term mobility could be communities "adapting" and developing resilience to "normal" floods (Penning-Rowsell et al. 2013: S45), and assistance programmes that follow floods, making migration unnecessary. Flooding, however, causes a lot of short-term population displacement, but few long-term moves (Gray and Mueller 2012), and evidence for long-term population relocation is inconclusive, with few large-scale quantitative studies (Penning-Rowsell et al. 2013). In the qualitative analysis, people talked about this temporary nature of flood-induced movements. The quantitative analysis shows that flooding has a negative influence on migration outside the district as well as long-term movement within the district. Migration does help people cope with climatic and environmental stresses and shocks; however, disasters can also sometimes reduce migration by cutting down on resources needed to migrate, or by increasing labour needs in the points of origin, as the literature shows (Gray and Mueller 2012;). The experience of flood, as observed in this book, seems to support what has been reported in the literature.

Yet another trend shown in the book is that of widespread experience of rainfall variability and its relationship with migration, as qualitative and quantitative analyses show. This observation chimes with rain gauge observations showing a hike in March–May rainfall by 3.4 per cent and a dip in June–August rainfall by 1.7 per cent between 1960 and 2003 (Karmalkar et al., n.d.). The literature also predicts heavier and more erratic rainfall in the whole of Ganga, Brahmaputra and Meghna system, with a wetter future for Bangladesh (World Bank 2010; Dasgupta et al. 2011; Immerzeel et al. 2013). Migration literature (Wolpert 1965, for instance) has identified rainfall variability as a key determinant of migration. Empirical literature – in the African context – shows that people from areas affected by rain-deficit and rainfall variability are more likely to leave their village compared with those living in wetter areas (Henry et al. 2004). The MARC model (Kniveton et al. 2011) explains how individuals make migration decisions when the rainfall patterns change (please see the literature reviewed in chapter 3).

EACH-FOR (2009) notes that households try to offset losses suffered in climate-related risks such as rainfall variability with short-term, seasonal or permanent migration to ensure better food and livelihood security. Following up on this research, Warner et al. (2013) reiterate that rainfall variability has an impact on household income and migration decisions. Dang (2014), however, has noted perceptions of climate variability as a factor that could also limit adaptation choices. However, Bohra-Mishra et al. (2014) report that in conditions that are initially dry, a reduction in rainfall increases migration, while in wetter conditions, an increase in rainfall increases migration. One of the logit models shows that a positive anomaly in rain increases long-distance migration, while normal rain reduces migration. It partially supports this observation of Bohra-Mishra et al. (2014). The rationale could be that in dry areas, an unexpected increase in rainfall could come as a boon, boosting crop production so long as it does not lead to floods and crop damage. On the contrary, in a wet place, such an anomaly could mean flood and crop damage. Negative anomaly in rainfall, however, does not appear to drive long-distance migration, as the literature shows. At the same time anomalies – negative as well as positive – and normal rain tend to be associated with house shifts within the district due to different reasons, as explained in chapter 5.

Frequent cyclones are considered a major driver of migration. Cyclone Aila, for instance, involved a storm surge that was four metres high along the coastal stretch, flooding areas up to 600 m away from the coast in low-lying areas, but spreading havoc on riverbanks and islands more than 40 km upstream along the river network (Gayathri et al. 2015). The lasting impact of cyclones includes failures in cropping and shrimp farming due to salinization, which could also alter migration patterns (WARPO 2006). Typically, after hazards, people move to safety and the landless among them move for income recovery (Penning-Rowsell et al. 2013). Besides, the latest census figures show a trend of migration away from the coast. The logit models show a clear association between long-distance migration and cyclone exposure.

In short, the book shows that climate- and environment-related stresses and shocks influence human movement is many different – and complex – ways. Studies have found that people often move for short periods over short distances as a result of disturbances to their habitats, safety and livelihoods (Poncelet 2008; Findlay and Geddes 2011; Gray and Mueller 2012;). However, such hazards often do not affect, or sometimes decrease, long-distance moves (Gray and Mueller 2012; Kniveton et al. 2008).

In that sense, this book only partially supports the finding of Penning-Rowsell et al. (2013) that a large share of migration is caused by riverbank erosion; and floods and cyclones do not cause permanent migraton unless they cause losses and long-term changes. While floods seem to discourage migration, as this book finds, cyclones and droughts influence migration outside the district. On the contrary, droughts and cyclones tend to negatively influence long-term movement within the district. At the same time, logit models show riverbank erosion having a significant positive influence on short-distance migration, in line with the literature. It displaces 50,000 to 200,000 people in Bangladesh every year (Mehedi 2010). As it destroys farms and homes (Zaman 1989), sometimes communities are displaced several times, in a dramatically different way compared with migration for economic gain (Hutton and Haque 2004). A study by Abrar and Azad (2004) in north-west Bangladesh found that on average households have been displaced 4.6 times by riverbank erosion. More recent research suggests that permanent migration is common among the people affected by erosion, even though the rate of erosion may have become more moderated (Penning-Rowsell et al. 2013). However, riverbank erosion has a negative influence on inter-district migration, as the quantitative analysis shows.

7.2.4 Immobility of people in hazard-prone places

Last, but not the least, the logit models also suggest that though migration is driven by poverty and income needs, the poorest, especially those without any assets, are often unable to migrate outside the district. This observation in the book is also in line with international literature that shows droughts – and famines – encourage migration by poor farmers, though not the poorest (Kniveton at al. 2009). This immobility is a cause for concern, as people who are unable to move out – the poorest and the most vulnerable – could get trapped in places exposed to climatic and environmental hazards. At the same time, the qualitative analysis shows that even while people migrate instinctively – in a planned manner – to escape from the ill effects of climate and environment on their livelihoods, they do not associate these movements to these elements. This apparent lack of appreciation of the real impact of climate and environment on their lives and livelihoods is possibly an issue that needs to be addressed, especially in the context of addressing immobility. People often do not appreciate risks involved in their livelihoods and environments, and such lack of risk appreciation could be detrimental to effective disaster risk reduction (Covello and Sandman 2001). This even leads to permanent migration to disaster-prone, vulnerable locations, a practice that is very common in coastal Bangladesh, as Paul and Rautray (2014) report. This apparent lack of appreciation of the real impact of climate and environment on their lives and livelihoods is possibly an issue that needs to be addressed.

The logit models indicate that experiences of certain hazards tend to negatively influence migration. This phenomenon requires a closer look, as long-distance migration is a selective process that favours younger persons with more education and assets in their search for better income. Logically that means poorer, older and less educated people might get left out of the process. The literature suggests that though migration to cities is a preferred option, not everybody has access to this mechanism, especially those with not many assets. It appears to be a paradox. While it is people in need of more income who migrate, those who cannot muster enough resources back home – those with no assets – tend to be left out of the migration process.

Flooding and riverbank erosion appear to have a negative impact of inter-district migration. As explained above, riverbank erosion tends to displace people more locally – still the concern is that the disaster tends to incapacitate people to move out of the district to earn a better income. Flooding is more of a case of temporary movement – until the rainy season ends and the floodwaters recede. People seasonally move back and forth from flood plains. However, such chronic exposure to floods appears to reduce the ability of people to seek long-term alternatives offered by inter-district or even short-distance migration.

Meanwhile droughts, flooding and cyclones appear to discourage movement within the district. This observation about drought and cyclones could have potentially serious implications for those with more assets migrating out of the district after exposure to these disasters; but those with less assets being unable to move even locally. In the case of flood, people do not appear to move anyway. It could be argued that with prolonged exposure to hazards, people might not be able to muster enough resources to undertake the journey – across short or long distances, as the case may be. That means vulnerable people could in effect be trapped in places exposed to climatic and environmental hazards.

There are several reasons behind people finding it difficult to migrate at times. Migration involves initial costs and social networks in the destination; so the loss of resources such as land back home and the uncertainties regarding the remuneration at the destination place huge demands on human, social and financial capital of the migrants that the poor among them cannot meet (Gray and Mueller 2012). The poorest people in villages often do not have enough resources needed for migration in terms of investments, networks and physical capability – in effect, their households getting trapped in vulnerable rural spaces (Mallick and Etzold 2015). Such people, especially the elderly and those in poor health among them, end up depending on locally available resources, including post-disaster aid and support from the home community, the long-drawn recovery process leading to dependency and indebtedness (Mallick and Etzold 2015). While mobility helps people cope with disasters and their long-term effects, the link between short-term post-disaster moves and permanent migration can be rather unclear and complicated (Mallick and Etzold 2015; Penning-Rowsell et al. 2013). Another reason that Black et al. (2014) suggest is that in the aftermath of disasters there could be changes in the local labour market that might encourage or discourage migration. Black et al. (2014) further argue that there could be reconstruction after disasters, leading to an economic buzz, creating jobs, benefitting local people, though not necessarily those affected by the disaster. Such reconstruction activity after Hurricane Katrina, for instance, attracted undocumented Hispanic migrant workers, who were made to work at lower wages compared with that of local workers (Fletcher et al. 2007).

One local example cited in this regard is that of floods in Bangladesh, contributing to better agricultural wage rates in the long term (Banerjee 2007). The need for replanting crops could open up the labour market, making migration less likely. On the contrary, environmental events could also lead to lower farm productivity or wages, forcing poor people to migrate for work. The qualitative analysis shows the trend of landless and marginal farmers migrating more than others in the aftermath of weather uncertainties and associated losses. At the same time, the quantitative analysis does not show any significant relationship between flood and migration. Possibly such a lack of connection could indicate different trends of mobility and immobility cancelling out one another.

The notion of people being trapped in hazard-prone places is a relatively new one in climate- and environment-related migration. It is not only resource constraints, socio-economic factors and policy barriers that limit adaptation choices, but also psychological factors, practices and perceptions of climate variability (Dang et al. 2014). Sometimes people just may not want to move despite the risks involved in staying at a vulnerable place (Black and Collyer 2014: 52). Sometimes people opt to stay back no matter what. In their Bangladesh study, Penning-Rowsell et al. (2013) argue that people are rooted to their places for their livelihood and housing, leaving only the landless and men without families able to move to work in farms elsewhere for daily wages. One example is the case of Gabura, where in an initial focus group discussion, the villagers described how Cyclone Aila in 2009 left the place inundated, salinised and thereby barren for years to come. Men had to migrate for many months. Still they preferred to live there as it was home (Martin 2015). Gardner (2009: 233) notes that people in Bangladesh in general have a strong sense of rootedness and home.

The sense of rootedness is characteristic of rural communities elsewhere as well. In the north-west Costa Rican context, Warner et al. (2015) show that farmers perceived rice production as an identity, and to maintain that they had to limit their adaptation choices. For households that did not have enough water, adaptation meant decreased rice-market access. They become trapped by their inability to reduce their vulnerability (Warner et al. 2015). As Collins (2013) argues, such slow-onset environmental crises become a part of life and people adapt to them in situ. Even when displacement is likely after a shock, accumulated impacts can produce "acceptance, adaptation and resilience" in situ until deterioration of assets severely limits even the option to stay back (Collins 2013: S114). Studying rainfall variability, food insecurity and migration in Guatemalan mountain communities, Milan and Ruano (2014) have reported on the local people's experiences of climatic conditions worsening, affecting food production in the last 20 years. With reducing options for local livelihood diversification and limited migration opportunities, they run the risk of becoming stuck in a place vulnerable to climate change. As Penning-Rowsell et al. (2013) have noted in the context of Bangladesh, despite the major hazards and their threats, permanent movement/migration is limited, and all movement is to some extent reluctant.

Whether people state it explicitly or not, migration can be interpreted as a strategy that helps them offset losses suffered in climatic and environmental stresses and shocks and be prepared better for future stresses and shocks. However, whether such migration leads to adaptation depends on the policy environment in the country. A textual analysis of relevant policy documents, however, shows that though urban migration is an inevitable part of Bangladesh's economic growth, its role as a climate change adaptation strategy or its links with hazards and local vulnerabilities are often not acknowledged at a policy level. The book argues that policies need to be more proactive so that migration does not become maladaptive or people unable to move out do not get trapped in places exposed to climate- and environment-related hazards.

7.3 Migration as adaptation

In Bangladesh's case, recent literature has reiterated that high levels of rural poverty coupled with rapid urban growth drives migration (Muzzini and Aparicio 2013; Marshall and Rahman 2013). Still such migration arguably has a climate and environment nexus, as changes and uncertainties affect livelihoods. In a country exposed to various hazards and extreme weather (Harmeling 2012), dubbed as a climate change hotspot (Huq 2001; Huq and Ayers 2008) with a high population density (BBS 201), widespread poverty (World Bank 2013) and a heavy dependence on natural resource-based primary livelihoods, human impacts of climatic stresses and shocks get amplified (Agrawala et al. 2003). To offset this impact, people living in climate-sensitive areas increasingly adopt secondary livelihoods that are not dependent directly on natural resources (Ahmad 2012), a trend that leads to an increase in urban migration (Afsar 2003; Muzzini and Aparicio 2013; Planning Commission 2011).

There are several examples of people using migration to offset these impacts of poverty. Villages affected by droughts, especially in northern Bangladesh, escape the lean period between harvests called *monga*, marked by poverty and food insecurity (Findlay and Geddes 2011). Landless labourers, mostly boys and men from *monga*-affected areas, often migrate to cities and better-off villages in search of work (Siddiqui 2009). While the climate- and environment-related migration in Bangladesh shows these characteristics described in the literature, it also demonstrates that networks play a key role (Bilsborrow and Okoth-Ogendo 1992; Munshi 2003; Schmidt-Verkerk 2011; Foresight 2011; Lu et al. 2012,; among others).

In places that experience high levels of climate-related mobility such as Bangladesh (Foresight 2011), migration becomes foremost a livelihood strategy amidst multiple opportunities, stresses, shocks and, above all, uncertainties – an activity interwoven with other societal processes (McLeman and Smit 2006). "If we assume that climate-stimulated migration is not simply a random or wholesale outpouring of people from an exposed area, migration can be seen as one possible manifestation or outcome of adaptive capacity in the light of exposure to some form of climatic stress" (McLeman and Smit 2006: 35). In such a scenario adaptation may be defined as an adjustment in ecological, social or economic systems as a response to observed or expected changes in climatic stimuli; and the effects and impacts of such stimuli to alleviate adverse impacts of change or take advantage of new opportunities (McCarthy et al. 2001; Adger et al. 2005).

Migration can be a good strategy for adaptation to environmental change, an "extremely effective" way to gain long-term resilience (Foresight 2011:10). For poor people, the lack of capacity to adapt to environmental risks or hazards may be interconnected with population displacements. As McLeman and Smit (2006) argue, the Dust Bowl migrants of rural Oklahoma in the 1930s were displaced due to the combined effect of crop failure following drought, land use, tenancy, social networks and ability of family members to migrate – factors that together influenced their adaptive capacity. Recent literature argues that far more than a coping strategy, migration can be a positive adaptation measure (Barnett & Webber 2010). Still it is often difficult to determine whether climate- and environment-related human mobility indicates successful adaptation or a failure to adapt in situ (Warner and Afifi 2014).

The qualitative analysis shows that villagers in districts exposed to climate- and environment-related stresses, shocks and uncertainties are diversifying their traditional livelihood strategies by migrating. Although the migrants' primary motivation is better livelihoods and better income, climatic and environmental factors work in the background, often influencing their migration decisions, or sometimes making migration necessary. As migration contributes to produce better and safer lives and livelihoods, especially in the context of changing or uncertain climatic factors, it may be argued that migration becomes an effective form of adaptation.

7.4 Scope for follow up studies

In a context of climate change, this book probes to what extent climate- and environment-related stresses and shocks influence villagers' decisions to stay put or move out of their place in Bangladesh. Based on neoclassical migration theories, influenced by social psychology and behavioural economics and informed by climate sciences, the study also includes a strong policy element with wider practical implications. While debating and negating some old notions of climate-induced mass migration largely based on neo-Malthusian economics, this piece of research underscores the multi-causal nature of migration. Even while migration is primarily driven by livelihood and income needs, climatic and environmental stresses and shocks influence people's migration decisions, even though people seldom attribute their movements to these factors. In that sense, it places migration as an effective adaptation strategy and looks at the broader policy implications of such a notion. In this context, the research potentially opens up opportunities for enquiries on several fronts.

Firstly, the research considers cognitive aspects of decision-making amidst uncertain and incomplete information on weather, climate and their impacts on the ground. The book acknowledges advances in climate modelling that predict a wetter future for the upstream areas of the three major rivers whose low-lying delta makes up much of Bangladesh – thereby making large parts of the country hazard-prone. It offers possibilities to study three dimensions of this phenomenon. It is important to understand the future climate of Bangladesh based on climate models and its impact on the hazard profile of the country mentioned in chapter 2; its potential repercussions on livelihoods and safety of people; and their influence on people's migration decisions and patterns.

Secondly, this book argues that villagers in Bangladesh migrate for better livelihoods and use migration as a response to climatic stresses and shocks as an adaptation or coping strategy. Most of the movements are reported to be within the country, and if facilitated appropriately, such internal migration – along with international migration – can be an effective adaptation strategy, as the book shows. Such movements can help build individuals' and communities' adaptive capacity to future environmental and climatic hazards. Migration happens in the context of a growing city-centred economy that promotes remittances to villages. An area that requires further studies is the adaptive nature of migration, the efficacy of migration as an adaptive strategy and factors that aid or inhibit movements of people.

Thirdly, future studies could probe how policies can help migration and boost its adaptation potential. A textual analysis of current and recent policies concerning climate change, development and poverty alleviation, and disaster management in Bangladesh shows that the economic and adaptive roles of internal migration are often not included in policy documents. More than that, as explained in the policy analysis part of the book, these policies tend to tacitly discourage migration even when an overall development strategy of city-based growth promotes rural-urban migration. This ambiguity and its impacts are something that require closer research attention.

Fourthly, a related theme for further enquiries is restrictions to mobility. The book shows that hazards and their aftermath could involve serious implications for people's mobility or, even more seriously, lack of it. It indicates that a large number of people might be trapped under risky circumstances, exposed to climate- and environment-related hazards. While it may be argued that internal and external factors could in effect trap people in vulnerable places, mobility could help them escape hazards and offset their impacts. Mobility involves the freedom to seek opportunities to improve livelihoods, living standards and services such as health care and education – succinctly put, safer and more productive lives in more responsive communities (UNDP 2010). Often disasters take away resources and assets that contribute to such mobility.

Further, from a local perspective, there is a need to understand the influence of climatic and environmental stimuli on people's links and rootedness with their place and probe into cognitive processes that define these links. The challenge here is to understand to what extent people stay put in a hazard-prone place and what it takes to move out; whether they are unable to move out, are they likely to get trapped; and to what extend uncertain and fuzzy elements of climate projections can be translated into actionable policy interventions. Such an in-depth knowledge would be beneficial to local communities trying to deal with hazards and policymakers addressing emerging challenges, concerns and opportunities. Particularly, the inevitable loss and damage – climate change effects that people cannot adapt to or cope with – is a matter or emerging policy interest (UNFCCC 2013). Immobility related to loss and damage, therefore, is a field for potential future enquiry.

Fifthly, a weakness in this book is that it does not look closely enough at short-term, short-distance moves, especially after hazards, which apparently constitute a major share of climate- and environment-related mobility, as the literature shows. The quantitative survey included questions on short-term moves, but the data gathered was not detailed enough for analysis. While it may be important to study the duration, distance and frequency of movements after these events, attention should also be paid to understanding the benefits and costs involved in these moves. Staying or moving – over different spans of time and space – could be a response mediated by hazard experiences and risk perceptions, as the qualitative study shows.

Overall, there is scope for interdisciplinary research, especially on extreme and uncertain climate and weather events and the possibility of people living in vulnerable settings – places and socio-economic circumstances – becoming unable to adapt or cope with the changes in their climate and environment. Climate and oceanographic models could shed more light on extreme precipitation, sea level rise and cyclone intensity, trajectories and storm surges, and contribute to research on their human impact. A step ahead, there could be closer examination of how effective risk appreciation and communication could be promoted under such circumstances.

Annexes

1 Qualitative questionnaire[1]

1.1 Village-level community survey

INTRODUCTORY INFORMATION

Name of the Village/ Ward No
Name of union
Name of thana
Name of district
GPS location
Any change in name or status of village since 1970? (Yes/No)

If changes occurred mention those:

1 Physical characteristics

1.1 Total number of population:
1.2 Total number of arable land of the village:
1.3 Total amount of non-arable land of the village:
1.4 Total number of Khas ponds/ditch/marshland/ in the village:
1.5 Educational qualification of the villagers:

Educational qualification	*Percentage (%)*
Educated	
Half-Educated	
Less- Educated	
Able to sign (literate)	
Illiterate	

1 This questionnaire was prepared by Dominic Kniveton, with inputs from CDKN project colleagues.

1.6 What are the sources of clean water?

Tube well	
Well	
Pond	
Rain water	
Others	

1.7 When scarcity of water is experienced by the villagers?
1.8 Number of landless people in the village:
1.9 What are the sources of cooking fuel in the village?

Firewood	
Paddy	
Cow-dunk/Muck	
Chaff/Husk of grain	
Jute-stalk	
Tree leafs	
Chaff wood	
Others	

1.1 Is there any government owned or private forest in the village? If so, per cent of total area (in square km)?

2 Educational facilities

2.1 How many primary schools are there in the village? List them:
2.2 Queries for each Primary School

No	*Name of the School*	*When did it open (if after 1970)? [year]*	*Type*	*Are there any years in which the school was temporarily closed? [In years]*	*Reason for closure?*
1					
2					
3					
4					
5					

Types

Bangla medium (Government)	1
Bangla medium (Non-government)	2
Madrasha	3
Kindergarten	4
English Medium	5
Non-formal/Vocational school	6

2.3 If there is no primary school, how far away is the nearest one? [km]
2.4 List all the secondary schools in the village:

2.4.1 Queries for each one:

No	*Name of the School*	*When did it open (if after 1970)? [year]*	*Type*	*Are there any years in which the school was temporarily closed? [In years]*	*Reason for closure?*
1					
2					
3					
4					
5					

Types

Bangla medium (Government)	1
Bangla medium (Non-government)	2
Madrasha	3
Kindergarten	4
English Medium	5
Non-formal/Vocational school	6

2.5 If there is no secondary school, how far away is the nearest one? [km]

3 Health facilities

3.1 List all clinics, dispensaries and hospitals in the village. Queries for each one:

No	*Name*	*When was it established (if after 1970)? [year]*	*facilities*	*Are there any years in which the facility was temporarily closed? [In years]*	*Reason for closure?*
1					
2					
3					
4					
5					

Clinics	1
Hospitals	2
Dispensaries	3
Homeopathies	4
Ayurvedic physician	5
Religious healers	6

3.2 List all pharmacies in the village. Queries for each one:

No	*Name of the pharmacy*	*When did it open (if after 1970)? [year]*	*facilities*	*Are there any years in which the facility was temporarily closed? [In years]*	*Reason for closure?*
1					
2					
3					
4					
5					

3.3 If there is no clinic, hospitals, pharmacy, dispensary, Ayurvedic physician, homeopathies in the village, how far away is the nearest one? [km]

4 Population

4.1 How many religious communities or ethnic groups live in the village?

List the 4 largest ethnic groups:

Communities/ ethnic groups	*Percentage (%)*	*Total Household*	*How many years are they living in the village?*	*Decrease (1) Increase (2)*
Muslims				
Hindus				
Christians				
Buddhists				
Ethnic				
Others				

4.2 Has any new community moved into the village?

4.2.1 Is there a period in which they came or left in large number? [Y/N]
4.2.2 When did this period did begin and end? [year]

5 Diseases

5.1 Please tell us about the major diseases that have affected the village since 1970.

No	*Name of diseases*	*At present which diseases are frequently occurring?*	*In which year did the disease appear?*	*Which diseases frequently occurred in 20 years before?*	*Year of disappearance of these diseases?*
1	Diarrhoea				
2	Hepatitis				
3	Arsenic				
4	Skin Diseases				
5	Dengue				
6	Kala-azar				
7	TB				
8	Malaria				
9	Leprosy				
10	Small pox				
11	Others				

6 Harvest failures

6.1 Are there any years since 1970 when the harvest has been particularly bad? [Y/N]

Ask in decade: (2012–2001); (2000–1990); (1989–1980); (1979–1971)

6.2 In which years were the harvest failures worst?

Year	*Reasons (Drought, Excessive rain, Shortage of rainfall, flooding, Storm/ cyclone, damage or attack by pests, river erosion others)*	*Did those affected sell livestock? [Y/N]*	*Did those affected sell possessions? [Y/N]*	*Did people leave the village? [Y/N]*	*Did people go hungry? [Y/N]*	*Did people adjust their diet? [Y/N] such as eating potato, Muri (fried rice), Chira instead of rice*	*Was there mutual assistance? [Y/N]*	*Other*

7 Principal economic activities

7.1 Please tell us about the principal economic activities in the village since 1970.

Profession	*Most (over 80%)*	*Many (over 50%)*	*Some (over 10%)*	*Hardly any (under 10%)*	*Before 1970*	*Year after 1970 (When?)*
Agriculture						
Fishing						
Shrimp culture						
Poultry						
Livestock						
Fish farming						
Craft production						
Commerce						
Industry/mining						
Services						
Plantation business						
Others						

7.2 How many crops are cultivated in a season in this village?

7.3 Please tell us about the principal crops grown in the village since 1970.

Crops name	*When did the production of the crop start? [year]*	*Is it mainly grown for consumption or sale, or both?*	*If mainly sold, when did it start to be sold? [year]*
Paddy			
Jute			
Wheat			
Pulse			
Potato			
Tobacco			
Vegetables			
Sugarcane			

7.4 Please tell us about the agricultural technology used in the village since 1970.

Agricultural technology	*Is this used? [Y/N]*	*When did the villagers start using it? [year]*	*Is this widely used? [Y/N]*	*When did it become widely used? [year]*
Tractors/Power Tiller				
Pump Irrigation				
Chemical fertilizers				
Threshing Paddy machine/ Grinding mill				
Rice processing machine				
Spices processing machine				
Hybrid Seeds				
Others				

7.5 Forms of land tenure found in the village?

Ownership-Possession/land tenure system	*Percentage (%) of own land*	*Farming in Rented or mortgaged land (%)*	*Sharecropping/ Tenants farming (%)*	*Mortgaged land (%)*
Present Situation				
20 years before				

7.6 Related to non-arable land:

Non-arable land	*Yes*	*No*	*Year*
Is there non-arable land in the village?			
Has there been any change since 1970?			
In which year were the changes observed?			
Did the size of non-arable land get larger or smaller?			

7.7 Is there any change in land use pattern?

Got Waterlogged	
Lack of irrigation water	
Intrusion of saline water	
Shrimp culture	
Homestead in the cultivable land	
Brickfield in the cultivable land	
others	

7.8 Has farming in the village been affected by the weed? (Yes/No)

7.8.1 When did the weed problem start? [year]
7.8.2 Has it been eradicated? [Y/N]
7.8.3 If Y, when? [year]

8 Employment

8.1 Is it possible to find paid work in the village?

8.1.1 Which sector?
8.1.2 What activity?

8.1.3 What is the name of the employer/company?

List each enterprise offering paid work:

Name of the employer/ company	*When did the enterprise start? [year]*	*When did the enterprise finish? [year]*	*How many are employed? [(>50) More than 50; 10–50; (10 to 50); (<10) less than 10]*	*Is employment temporary/ seasonal/ permanent? [Y/N]*	*Have people from other villages moved to the village to take up this employment?*

8.2 Are there any sources of paid work in the village that have existed since 1970, but no longer exist? List each enterprise that offered paid work:

8.2.1 Which sector?
8.2.2 What activity?
8.2.3 What was the name of the employer/company?

Name of the employer/ company	*When did the enterprise start? [year]*	*When did the enterprise finish? (year)*	*How many are employed? [(>50) More than 50; 10–50; (10 to 50); (<10) less than 10]*	*Is employment temporary/ seasonal/ permanent? [Y/N]*	*Have people from other villages moved to the village to take up this employment?*

8.3 Is it possible to find paid work close to the village? List each enterprise offering paid work:

8.3.1 Which sector?
8.3.2 What activity?
8.3.3 What is the name of the employer/company?

Name of the employer/ company	*When did the enterprise start? [year]*	*How far away is the enterprise? [km]*	*How many are employed? [(>50) More than 50; 10–50; (10 to 50); (<10) less than 10]*	*Is employment temporary/ seasonal/ permanent? [Y/N]*	*Have people from other villages moved to the village to take up this employment?*

8.4 Are there any sources of paid work close to the village that existed after 1970, but no longer exist? List each enterprise that offered paid work:

8.4.1 Which sector?
8.4.2 What activity?

Name of the employer/ company	*When did the enterprise start? [year] if abolished, which year?*	*How far away is the enterprise? [km]*	*How many are employed? [(>50) More than 50; 10–50; (10 to 50); (<10) less than 10]*	*Is employment temporary/ seasonal/ permanent? [Y/N]*	*Have people from other villages moved to the village to take up this employment?*

8.4.3 Major professions (Past and Present)

Profession	*Existed before (20 years) but no more*	*Present/Still exists*
Day labour		
Shopkeeper		
Fisherman		
Rickshaw/Van puller		
Driver of auto (C.N.G., Scooter, Bus, Truck)		
Boatman		
Confectioner/sweet maker		
Milkman		
Potter		
Blacksmith		
Weaver		
Ghee seller		
Oil seller (ferry)		
Scrape collector		
Mason		
Carpenter		
Mobile phone Booth, Flexi load business		
Honey collector		
Teacher		
Farmer		
Share cropper		

9 Development projects and associations

9.1 How many development projects are currently operating in the village?

For each project:

Name of the project (Government/Non-government/ N.G.O. operated). If Govt. Project, name of ministry/ Department.	*When did it start? [year]*	*What sector? [agriculture, livestock, credit, health, education, other]*

9.2 Are there any development projects in the village since 1970 that are no longer operating? [Y/N]

Name of the project (Government/Non-government/ N.G.O. operated) If Govt. Project, name of ministry/Department.	*When did it start? [year]*	*When did it finish? [year]*	*What sector? [agriculture, livestock, credit, health, education, other]*

9.3 Are there any village associations or co-operatives? [Y/N]

For each association:

Name of the associations or co-operatives.	*When did it start? [year]*	*Is it for men, women, or both?*	*What sector? [agriculture, livestock, credit, health, education, other]*

9.4 Are there any village associations that used to exist since 1970, but no longer exist? [Y/N]

For each association:

Name of the associations or co-operatives.	*When did it start? [year]*	*When did it finish? [year]*	*Is it for men, women, or both?*	*What sector? [agriculture, livestock, credit, health, education, other]*

9.5 Are there any Diaspora associations or associations of migrants? [Y/N]

For each association:

Name of the association	*Which country/Where is it based?*	*When did it start? [year]*	*What sector? [cultural activities; economic support to the village; integration of migrants at destination]*

9.6 Are there any diaspora associations or associations of migrants that used to exist since 1970 but no longer exist? [Y/N]

For each association:

Name of the association	*Which country/ Where is it based?*	*When did it start? [year]*	*When did it finish? [year]*	*What sector? [cultural activities; economic support to the village; integration of migrants at destination]*

10 Village infrastructure

10.1 Do you have the following in the village?

Name	*Yes*	*No*	*If No, how far away is the nearest place you can find it? [km]*	*If Yes, since when has it been available? [year]*
Grazing land				
Safe drinking water supply				
Piped drinking water				
Irrigation system				
Grinding mill				
Cinema				
Credit institution				
Electricity				

10.2 Do you have a market in the village? [Y/N]
10.3 When did the local market first start? [year]
10.4 Is the market daily or weekly?
10.5 Has the frequency changed? [year]
10.6 Can you buy the following in the village?

Goods	*Yes*	*No*	*If No, how far away is the nearest place you can find it? [km]*	*If Yes, since when has it been available? [year]*
Cooking oil				
Salt				
Sugar				
Soap				
Gasoline				
Batteries				

10.7 Does the village have a paved road? [Y/N]

10.7.1 If N, how far away is the nearest one? [km]
10.7.2 If Y, when was it built? [year]

10.8 Does the village have motorized public transport? [Y/N]

10.8.1 If N, how far away is the nearest place? [km]
10.8.2 If Y, when was it first available? [year]

10.9 Does the village have a boat landing? [Y/N]

10.9.1 If N, how far away is the nearest place? [km]
10.9.2 If Y, when was it first available? [Year]

1.2 Focus group discussion

VILLAGE:

UNION:

DISTRICT:

LIVELIHOOD

Family	
Upper class	
Middle class	
Lower class	

1.1 What changes have occurred in the livelihood of your family in the last 40 years?
1.2 How many livelihoods were there and what were these?
1.3 What did your ancestors (father, grandfather) do?
1.4 What was your occupation before and what are you doing now?
1.5 What are the common problems of this village in perspective of livelihood, which are not related to environment and climate change?

No	*Livelihood*	*Common problems*

2

2.1 What are the common problems of the people in this village? (Except Environment).
2.2 Have the villagers ever thought of migration to solve any of these problems?
2.3 What kinds of changes have taken place in the climate in this area over the last 40 years? What have you heard from your parents, grandparents and neighbours?
2.4 Which was the worst natural calamity people have ever seen and/or faced in their life? (Area specific)

3 The chronological events that occurred due to climate change in the past:

No	*Name of the incidents*	*Year*	*Time/Month*	*Death incidentals*	*Damage of individual property*	*Losses of government/ non-government infrastructure*

3.1 We heard about workable indigenous knowledge with regard to weather forecasting? What were/are those?
3.2 Does indigenous knowledge work now? (Y/N)
3.3 What sort of thing do you find different from indigenous knowledge?

4 What are the major climate-change-induced events the upper/middle/lower class family faces now?

No	*Family*	*Impacts/Result and responses*
	Upper class	
	Middle class	
	Lower class	

4.1 Generally what sort of natural calamities do people face?

Please explain the variability of climate change

Rainfall

Quantity/Degree/intensity	
Duration	
Present intense months	
Past intense months	
Problems faced in agriculture production/cultivation	
Changes needed to bring in cultivation	
Problems faced in other occupation except agriculture	
What steps have people taken to deal with this problem?	

Warming

Quantity/Degree/intensity	
Duration	
Present intense months	
Past intense months	
Problems faced in agriculture production/ cultivation	
Changes needed to bring in cultivation	
Problems faced in other occupation except agriculture	
What steps have people taken to deal with this problem?	

Drought

Quantity/Degree/intensity	
Duration	
Present intense months	
Past intense months	
Problems faced in agriculture production/ cultivation	
Changes needed to bring in cultivation	
Problems faced in other occupation except agriculture	
What steps have people taken to deal with this problem?	

Cyclone

Quantity/Degree/intensity	
Duration	
Present intense months	
Past intense months	
Problems faced in agriculture production/ cultivation	

Quantity/Degree/intensity	
Changes needed to bring in cultivation	
Problems faced in other occupation except agriculture	
What steps have people taken to deal with this problem?	

Flood

Quantity/Degree/intensity	
Duration	
Present intense months	
Past intense months	
Problems faced in agriculture production/ cultivation	
Changes needed to bring in cultivation	
Problems faced in other occupation except agriculture	
What steps have people taken to deal with this problem?	

Riverbank Erosion

Quantity/Degree/intensity	
Duration	
Present intense months	
Past intense months	
Problems faced in agriculture production/ cultivation	
Changes needed to bring in cultivation	
Problems faced in other occupation except agriculture	
What steps have you taken to deal with this problem?	

4.2 Can you rate people's current situation with regards to your resources to carry out different livelihood activities? (go through different livelihood responses).

4.3 Please rate how threatening the following problems are to the sustainability of people's livelihoods:

Problems	*Not at all threatening (1)*	*Little threatening (2)*	*Threatening (3)*	*Threatening to a great extent (4)*	*Extremely Threatening (5)*
Flooding					
Drought					
Rainfall					
Increased temperature					
Riverbank erosion					
Water stress of plants					
Salinisation of water resources					
Salinisation of soil					
Increased pests					
Increased disease load					
Erosion of soil					

4.4 If need be, please elaborate how people's livelihood has been affected by rainfall/warming/drought/cyclone/riverbank erosion?

4.5 Please rate how likely people think the following problems are to occur in the next 10, 20, 30, 50 years:

Problems	*Not at all (1)*	*Sometimes (2)*	*(3) Regular Intervals*	*(4) Frequent*	*All the time (5)*
Crop failure due to flooding					
Crop failure due to drought					
Complete salinity of water resources					
Complete salinity of soil					
Loss of livelihood potential due to loss of land to erosion					
Inability of getting produce to market due to flooding					

4.6 What are the problems people face to sustain livelihood?
4.7 What changes have people taken so far in their livelihood due to climate-change-induced events?

Changes of job: temporary/permanent	
Profit/loss due to change in occupation	
Increase capacity of occupation by following adaptation strategy	
What kind of adaptation strategy you follow	
How effective are they (adaptation strategy)?	
How costly (economic and social) are they?	

5 What are the general problems people face in this locality due to climate-change-induced events?

Scarcity of drinking water	
Insecurity of women	
Irregular or stagnated academic curriculum in educational institutions	
Water logging	
Intensification or spread of diseases	
Infrastructural damage	
Others	

6 What are the causes of people's migration from this area? How has migration been influenced by climatic events?

1.3 Individual semi-structured interview

1 Basic Household Information

Serial No	*Relationship with the Household head*	*Age*	*Education*	*Female 1 Male 2*	*Income source*	*Monthly Income*
1	H Head					
2						
3						
4						

(*Continued*)

Serial No	*Relationship with the Household head*	*Age*	*Education*	*Female 1 Male 2*	*Income source*	*Monthly Income*
5						
6						
7						
8						
9						
10						
Total Number						Total Family Income =

If any family member is migrant then put an asterisk (*) beside him/her.

Code

Relationship with the Household Head	*Education*
1 Wife	1 Not Literate
2 Husband	2 Can sign
3 Daughter	3 Class 1 to 5
4 Son	4 Class 1 to 5
5 Mother	5 SSC/Dakhil/O' Level/ Vocational SSC
6 Father	6 HSC/Alim/A' Level
7 Sister	
8 Brother	

2 Land Holdings Description

Homeland	*Agricultural Land*	*Non-agricultural*	*Pond*

2.1 What is the portion at the pond in case of joint ownership?
2.2 Other Properties apart from land.

Properties	*Quantity/Number*	*Value*
Shop Owner		
Business in leasing shop		
Rice Mill/Threshing Machine		
Irrigation Pump		
Power Tiller		
Chicken Farm		
Business of agricultural Product (Seeds, Fertilizer)		
Fish Trader		
Fish Farm		
Trees/Fruits Orchard		
Cattle Farm		
Others		

3 Migrants' Profile

Serial No	*Destination*	*Years of Migration (staying period)*	*Cost*	*Types of Job*	*Skill*

4 Livelihood

4.1 What changes have occurred in the livelihood of your family in the last 40 years?

4.1.1 How many livelihoods were there and what were these?
4.1.2 What did your ancestors (father, grandfather) do?
4.1.3 What was your occupation before and what are you doing now?

4.2 Were your family members indigenous/inhabitants of this area or migrants?

4.2.1 How long ago did they come?
4.2.2 From where did they come?

5 Cognitive

5.1 What are the common problems of your family?
5.1.1 Have you ever thought of migration to solve any of these problems?
5.2 What kinds of changes have taken place in the climate in your area over the last 40 years? What have you heard from your parents, grandparents and neighbours?

5.2.1 Which was the worst natural calamity you have ever seen and/or faced in your life? (Area specific)
5.2.2 We heard about workable indigenous knowledge with regard to weather forecasting? What were/are those?
5.2.3 Does indigenous knowledge work now? (Y/N)
5.2.4 What sort of thing do you find different from indigenous knowledge?

5.3 What are the major climate-change-induced events you or your family face now?

Please explain the variability of climate change:

Rainfall

Quantity/Degree/intensity	
Duration	
Present intense months	
Past intense months	
Problems faced in agriculture production/cultivation	
Changes needed to bring in cultivation	
Problems faced in other occupation except agriculture	
What steps have you taken to deal with this problem?	

Warming

Quantity/Degree/intensity	
Duration	
Present intense months	
Past intense months	
Problems faced in agriculture production/cultivation	
Changes needed to bring in cultivation	
Problems faced in other occupation except agriculture	
What steps have you taken to deal with this problem?	

Drought

Quantity/Degree/intensity	
Duration	
Present intense months	
Past intense months	
Problems faced in agriculture production/cultivation	
Changes needed to bring in cultivation	
Problems faced in other occupation except agriculture	
What steps have you taken to deal with this problem?	

Cyclone

Quantity/Degree/intensity	
Duration	
Present intense months	
Past intense months	
Problems faced in agriculture production/cultivation	
Changes needed to bring in cultivation	
Problems faced in other occupation except agriculture	
What steps have you taken to deal with this problem?	

Flood

Quantity/Degree/intensity	
Duration	
Present intense months	
Past intense months	
Problems faced in agriculture production/cultivation	
Changes needed to bring in cultivation	
Problems faced in other occupation except agriculture	
What steps have you taken to deal with this problem?	

Riverbank Erosion

Quantity/Degree/intensity	
Duration	
Present intense months	
Past intense months	
Problems faced in agriculture production/ cultivation	
Changes needed to bring in cultivation	
Problems faced in other occupation except agriculture	
What steps have you taken to deal with this problem?	

5.4 Can you rate your current situation with regards to your resources to carry out different livelihood activities? (go through different livelihood responses)

5.5 Please rate how threatening the following problems are to the sustainability of your livelihoods:

Problems	*Not at all threatening (1)*	*Little threatening (2)*	*Threatening (3)*	*Threatening To a great extent (4)*	*Extremely Threatening (5)*
Flooding					
Drought					
Rainfall					
Increased temperature					
Riverbank erosion					
Water stress of plants					
Salinisation of water resources					
Salinisation of soil					
Increased pests					
Increased disease load					
Erosion of soil					

5.6 How satisfied are you with the difference between your current situation and your ideal situation with regards to resources?

Not at all satisfied	1
Little satisfied	2
Satisfied	3
Satisfied to a great extent	4
Extremely satisfied	5

5.5 If need be, please elaborate how your livelihood has been affected by rainfall/warming/drought/cyclone/riverbank erosion?

5.6 Please rate how likely you think the following problems are to occur in the next 10, 20, 30, 50 years:

Problems	*Not at all (1)*	*Sometimes (2)*	*Regular Intervals (3)*	*Frequent (4)*	*All the time (5)*
Crop failure due to flooding					
Cop failure due to drought					
Complete salinisation of water resources					
Complete salinisation of soil					
Loss of livelihood potential due to loss of land to erosion					
Inability of getting produce to market due to flooding					

5.4.1 What are the problems you face to sustain livelihood?

5.4.2 What changes have you taken so far in your livelihood due to climate-change-induced events?

Changes of job: temporary/permanent	
Profit/loss due to change in occupation	
Increased capacity of occupation by following adaptation strategy	
What kind of adaptation strategy did you follow?	
How effective are they (adaptation strategy)?	
How costly (economic and social) are they?	

6 What are the general problems you face in your locality due to climate-change-induced events?

Scarcity of drinking water	
Insecurity of women	
Irregular or stagnated academic curriculum in educational institutions	
Water logging	
Intensification or spread of diseases	
Infrastructural damage	
Others	

7 Causes of Migration and decision-making

7.1 Why did you or your family members migrate? Were migrations influenced by any climate-change-induced events (rainfall, temperature, drought, flood, cyclone)? Please elaborate.
7.2 What were the things or factors you or your family thought of before migration?
7.3 Who provided information about migration to you or your family?
7.4 What kind information was available to you before migration?
7.5 Who decided in family about member/s' migration?

Serial No	*Father*	*Mother*	*Brother*	*Sister*	*Self*	*Wife*	*Sister in law*	*Brother in law*	*Family*	*Others*

7.6 Who else (relatives, neighbours, friends) played a facilitating role in migration?

7.6.1 How did he/she play a role? Please elaborate or choose from below:

By giving money	
By providing information	
By giving job	
By giving information and job both	
By giving mental support	
By giving information on skill upgrading	
By supporting skill upgrading	
By providing information and support both for skill upgrading	
Others	

7.6.2 Did community play any role behind migration?

7.7 Why the whole family did not migrate?

8 Locus of Control

The fate of local people is in the hands of the people in power and there's not much that individuals can do about it.	*1*	*2*	*3*	*4*	*5*
The success of household is mostly determined by factors outside of your control.					
The weather and commodity prices can make life difficult in the short term, but in the long run there is still a lot you can do to stay ahead of the game.					
Many times I feel that I have little influence over the things that happen to me.					
No matter what things I try to make a living, the drought/flooding etc. prevents them from working.					

1 Strongly disagree; 2 Disagree; 3 Neutral; 4 Agree; 5 Strongly agree

9 Self-concept:

9.1 How often do people come to you for livelihood (including migration) advice?

Never	1
???	2
Sometimes	3
Frequent	4
All the time	5

9.2 How much influence do you think you have on other individuals when it comes to livelihood practices (including migration)?

No influence	1
Very Little influence	2
Little	3
Moderate influence	4
Significant influence	5

10 Risk and innovation:

10.1 Are you always one of the first in your area to change your livelihood in instances of stress and shocks? If yes, why?

If no, please elaborate.

10.2 Are you trying new livelihoods (such as migration) that aren't used a lot?

10.3 Are you willing to take more risks than other farmers in the area with respect to your production methods? If yes, why?

If no, please elaborate.

11 Trust Advice

11.1 For different livelihood responses: How much would you trust information on agricultural management from the following sources?

Institutions	*1 No trust at all*	*2 Little trust*	*3 Moderate trust*	*4 Trust*	*5 Complete trust*
Fellow households					
Local leaders					
Industry groups					
Agronomists					
Agribusiness					

Institutions	*1 No trust at all*	*2 Little trust*	*3 Moderate trust*	*4 Trust*	*5 Complete trust*
NGOs					
National Government					
Local Government					
Contractor					

11.2 To what extent is your decision to change livelihoods influenced by the behaviour of your neighbouring households, friends, family?

No influence	1
Very Little influence	2
Little influence	3
Moderate influence	4
Significant influence	5

2 Quantitative questionnaire[2]

SURVEY ON MIGRATION AND ENVIRONMENTAL CHANGE IN BANGLADESH 2012

Biographic Questionnaire
RMMRU/University of Sussex

IDENTIFIER: Household |__|__|__| Individual: |__|__|__|	**FAST READING:** **DATE :** |__|__| |__|__| |__|__|__|__|
VILLAGE/WARD: ____________	**IN-DEPTH READING:** **DATE :** |__|__| |__|__| |__|__|__|__|
	CODIFICATION : **DATE :** |__|__| |__|__| |__|__|__|__|
UNION: ____________	**DATA ENTRY:** **DATE :** |__|__| |__|__| |__|__|__|__|
THANA: ____________	**COHERENCE TESTS/CORRECTIONS:** **DATE :** |__|__| |__|__| |__|__|__|__|
	DATE : |__|__| |__|__| |__|__|__|__|
DISTRICT: ____________	**DATE :** |__|__| |__|__| |__|__|__|__|
	DATE : |__|__| |__|__| |__|__|__|__|
	INTERVIEWER'S NAME: ____________ **NO.** |__|__| **DATE :** |__|__| |__|__| |__|__|__|__|

2 This questionnaire was prepared by Richard Black, with inputs on climatic and environmental aspects by the author.

STARTING TIME _ _ hr _ _ min

PRESENTATION OF THE STUDY TO THE RESPONDENTS

Hello, my name is. . . . . I am taking part in a study on climate change-related migration in Bangladesh.

Before we start, I would like to briefly present the study to you and inform you about your rights.

This study has been organised by the Refugee and Migratory Movements Research Unit at the University of Dhaka and the University of Sussex in Britain.

WHY THIS STUDY? →

- The main aim of this study is to strengthen the ability of the Government of Bangladesh to understand, ***plan for and respond to migration in the future***, whilst reducing vulnerability, and building the resilience of the people of Bangladesh to the impacts of climate change.
- The ***dimensions and causes*** of migration remain poorly understood, including the extent to which it is influenced by climate change.
- The ***impact of migration*** on the living conditions of families or on the country's development has also not yet been adequately evaluated.
- With this study, which is organised by Bangaldeshi and British researchers, we would like to ***produce statistical data*** based upon the real life experiences of Bangladeshi people.
- The findings of this study will also be discussed on the occasion of ***public debates***, bringing together citizens, researchers and political decision-makers. The ultimwate objective of this study is hence to establish the link between real-life experience and migration and development policies.

HOW IS THIS GOING TO HAPPEN?

- To carry out this study, ***we meet people with very different migration experiences***: people who have always lived in this village, people who have lived in other places and who have returned, and also people who live currently away.
- A qualitative study has already been initiated, and has informed the development of this survey.
- Today, this survey concerns you, ***your life***. There are questions about the places where you lived since your childhood, about your occupations, your family life, the trips and stays you may have experienced, etc. All these questions will allow us to study the relationship between the **fact of migrating or not migrating** and the conditions in which you live.
- Since everyone's experience is different, ***the duration*** of the interview varies between 30 minutes and one hour and a half, depending on the person.

RIGHTS AND CONFIDENTIALITY OF ANSWERS

- If a question makes you feel uneasy, ***you are not obliged to answer***. You can stop the interview at any time.
- We guarantee that all the information you give us will be kept **confidential.**
- Your name and your address are not be recorded in the questionnaire – except your first name, so nobody will be able to identify you from the information that you give me.

Before we start: **Do you have any questions?**

Once any questions are answered. . . . Is it OK for me to continue with the interview?

INTRODUCTION — COLUMN 1

Q0 – First name of the respondent: . . . .

Q1 – The respondent is: 1. A man **2.** A woman

To begin with, I would like to note the major events and periods of your life on this grid. We will begin at the time of your birth. Later we will look at these periods of life in some more detail.

Q1A – To start with, and to help us to set a time scale, could you tell me in which year you were born (or your age)? |_1_|_9_|__|__|

GRID: Locate year of birth in the grid, note 0 in the columns titled "age," and fill in these columns by retracing the age till the current age of the respondent.

HISTORY OF HOUSING LIVED IN FOR AT LEAST 1 YEAR — COLUMNS 3.1 AND 3.2

Now we will talk about each *HOUSE OR APARTMENT* in which you lived for at least one year, starting from your childhood till now. We are interested also in moves you have made within the same town or village. . .

"House or apartment" includes all kinds of accommodation, including rented rooms, stays in hostels, staying with family members, etc. Note housing periods of at least one year in the grid. If there is space, note shorter periods in the "comments and specifications" column, e.g., "living for 5 months with uncle in Dhaka."

- *1st house:* **When you were born, in which town/village did you live? In which district?**

  GRID: at age 0, note in CAPITAL LETTERS the name of the town/village and of the district where the 1st house was located.

  Until when (what age) did you stay in this house?

  GRID: Locate the year of housing change and draw an arrow indicating the time spent in the first house.

- *2nd house:* **And then, where did you live? And until when did you stay in this house?**

  GRID: Note in CAPITAL LETTERS the name of the town/village and the district where the 2nd house was located and the year of moving into this new place. Ascertain the time spent in this second house and draw an arrow up to the 3rd house. . .

- *PROCEED in this way for each house until the current house and* ***go to Q2.***

Q2 – Is there a place which you would consider to be your village or your town of origin in Bangladesh? **1.** Yes
2. No ➔ *Q3*

Q2V – What is the name of this locality? . . . .

Q2D – In which district is it located? |__|__| *see list of district codes*

Q3 – Which group would you say you belong to in Bangladeshi society?

0. None **1.** Muslim **2.** Hindu **3.** Christian **4.** Buddhist **5.** Ethnic
6. Other, *Specify*: . . . .

CITIZENSHIP **COLUMN 9**

- **What is/are your nationality/nationalities by birth? Note the nationality or nationalities at** *year 0 in the column 9: Bangladeshi, Indian, etc. . . .*
- **And later on, did you change your nationality or acquire a new nationality/citizenship?** **1.** Yes
  2. No ➔ ***Family History: Q4***

  When did you change your nationality/citizenship?

  And which nationality/nationalities do you have at the moment?
  GRID: Note new nationality at every change that occurred. If multiple nationalities, note both/all the nationalities that the interviewee has.
- *Note: citizenship defined as passport(s) held/legal document(s) held.*

FAMILY HISTORY: PARENTS, BROTHERS AND SISTERS

Now let's talk about your family. . .

Q4 - How many brothers have you had in total? |__|__|
Take into account ALL brothers, even if they are not from the same fathers or mothers and even if they are deceased.

Q5 - And how many sisters have you had in total? |__|__|
Take into account ALL sisters, even if they are not from the same fathers or mothers and even if they are deceased.

Q6 - Are you the first-born of the family? **1.** Yes **2.** No

Q7 - Did your father work when you were 15 years old? **1.** Yes **2.** No **9.** DK **0.** Father unknown or deceased at that age ***if 2, 9,0 → Q10***

Q8 - Would you say he was:
Wage-earner: **1.** Higher-level occupation **2.** Skilled employee or worker **3.** Unskilled employee, worker, labourer
Non-wage employment: **4.** Employer **5.** Self-employed (without employees) **6.** Apprentice/trainee, intern **7.** Family help **9.** DK

Q9 - What was your father's level of education?
1. No schooling **2.** Junior school certificate **3.** Secondary School certificate **4.** Higher Secondary certificate **5.** Degree-level

Q10 - What was or were his nationalities?

1. . . . . **2.** . . . . **3.** . . . .

Q11 - Is he still alive? **1.** Yes **2.** No **→ Q11A – In which year did he die?** |__|__|__|__|

Q12 - And what was your mother's level of education?
1. No schooling **2.** Junior school certificate **3.** Secondary School certificate **4.** Higher Secondary certificate **5.** Degree-level

Q13 - What was or were her nationalities?

1. . . . . **2.** . . . . **3.** . . . .

Q14 - Is she still alive? **1.** Yes **2.** No **→ Q14A – In which year did she die?** |__|__|__|__|

FAMILY HISTORY: CHILDREN AND PARTNERS — COLUMNS 2.1 AND 2.2

We will now recollect the main events of your family life: the relationships, the children that you have had. Certain situations may not fit your personal life history, but this study has to be applicable to everybody, and we must therefore foresee all possible situations.

To begin let's talk about the *PARTNERS* that you have had in your life, whether you were married to them or not. Please indicate also those partners from whom you have separated or who are deceased.

- *1st relationship:* **When did your first relationship start?**
  To make it easier to remember, could you give me the first name of this person?
  GRID: Note: P (partner), the number of the partner and the first name of the partner – "P 1 Fatima" – in the grid at the start year of the relationship.
  Is this relationship still continuing today?
  If not: **When and how did it come to an end?**
  Note: S (separation), D (divorce) or DT (death) + no. of the partner + first name of the partner at the end year of the relationship: "D 1 Fatima"

  Did you have any *CHILDREN* with this partner? Please indicate also the children who are deceased.
  If yes: **In what year was the 1st child that you had with this partner born? (How old is the first child that you had with this partner now?)**

  What is his/her first name?
  GRID: Note in the grid: B (birth), the number of the child, the number of the relationship in which it was born and the first name of the child: "B1 P1 Sajida" at the year of birth.

  And the 2nd child that you had with this partner, when was he/she born?
  Note the birth of the 2nd child in this relationship at the year of birth in the grid, "B2 P1 Rasheed."

  And the 3rd child. . . *PROCEED in this way for all children born in this relationship.*

- **Have you had another relationship?** . . . *REPEAT the questions for each relationship: beginning and (possibly) end of every relationship + births and deaths of children.*

- *At the end: RECAPITULATE*: **Have you had any other children** *(outside of a relationship)* **?** **1.** Yes **2.** No
  GRID: Note these possible births outside of relationship in the grid: "B_OR Nazneen"

  Are all your children still alive? **1.** Yes **2.** No
  GRID: Note the death(s): DT (death) + the number of the child + the number of the relationship in which it was born + first name of the deceased child at the year of death: e.g.: "DT2 P1 Abdul."

IF EGO HAS NEVER HAD ANY PARTNER OR CHILD, MARK OFF: ➔ *NO PARTNER (Q15)* ➔ *NO CHILD (Q16)*

MIGRATION OF FAMILY MEMBERS AND PERSONAL NETWORK — COLUMN 4

Now I would like to talk with you about the places where your family members or other close relatives and friends have lived. . .

Since you were born:

Q15F - Has your **father** already lived for at least a year outside this District? **1.** Yes **2.** No *If yes: note 01* |__|__|

Q15M - And has your **mother** already lived for at least a year outside this District? **1.** Yes **2.** No *If yes: note 01* |__|__|

Q15B - And one or several of your **brothers**, have they already lived for at least a year outside this District?

0. No brother **1.** Yes **2.** No *If yes:* **Q15nB – How many?** |__|__|

Q15S - And one or several of your **sisters**? **0.** No sister **1.** Yes **2.** No *If yes:* **Q15nS – How many?** |__|__|

Q15P - And one of your **partners or previous partners**? **0.** No partner **1.** Yes **2.** No *If yes:* **Q15nP – How many?** |__|__|

Q15C - And one or several of your **children**? **0.** No child **1.** Yes **2.** No *If yes:* **Q15nC – How many?** |__|__|

Q15O - And **other relatives or close friends** you could count on, or could have counted on you to take in and to help you migrate to another place? **1.** Yes **2.** No *If yes:* **Q15nO – How many?** |__|__|

Q15T – Total : |__|__|

FILTER:* - *IF Q15T = 0 (No member of the family or personal network lived outside District for at least one year) ➔ ***Go to RELATIONSHIPS MODULE Page 10***

- Otherwise* ➔ *Describe the trajectory of each person. NEXT PAGE

TRAJECTORIES OF THE MIGRANTS AMONG FAMILY MEMBERS AND CLOSE FRIENDS AND RELATIVES

- *1st person: GRID, Note at the bottom of column 4:*
    - the **sex** of the person
    - the **first name** of the person (optional)
    - the **relationship** between this person and the respondent
      – Identify clearly the type of relationship:
      For the partners and children, record the identifier from the family-related columns 2.1 and 2.2 (e.g. P1, B4, etc.)
      For other persons indicate clearly: uncle, cousin, school friend . . .
    - *If the person is a friend or a partner:* **Since when do you know this person?** *Note the year at the bottom of the column..*

1st District: **Where was he or she living when you first met them?**

When did he or she start living there?
And until when did he/she live there? *Draw an arrow to indicate the period of time spent in the District.*

2nd District: **And then, in which District did he/she live for at least one year?** *Note the 2nd District at the year when the person started living there.*
Until when did he/she stay there? *Draw an arrow to indicate the period of time spent in this place.*

CONTINUE until the current place of residence and draw a line until today.

- *PROCEED in the same way with the second person. . .*

ATTENTION: *- Start the trajectories as early as possible. . . at least from the moment when Ego first met the person whose trajectory he/she is describing.*
- If the person is deceased: Note DT at the corresponding year.

MODULE: RELATIONSHIPS

I would now like to ask you some more questions about your relationships.
100 – Count in GRID (column 2.1) the number of relationships. You have had |___|___| relationships.

Questions	P 01	P 02	P 03	P 04	P 05
101 – No. of the partner/spouse *see GRID*	|__|__|	|__|__|	|__|__|	|__|__|	|__|__|
102S – 102E Start and end years of the relationship *See GRID – If ongoing cross out the end year*	Start End	Start End	Start End	Start End	Start End
If the relationship ended **103 – Type of dissolution** *(see GRID)* **1.** Separation or divorce **2.** Partner deceased	|__|	|__|	|__|	|__|	|__|
First name of the partner/spouse *Information not retained in data entry*	. . . .	. . . .	. . . .	. . . .	. . . .
104 – **What was his/her level of education at that time?** **1.** No schooling **2.** Completed Junior school **3.** Completed Secondary school **4.** Completed Higher Secondary **5.** Completed Degree-level	|__|	|__|	|__|	|__|	|__|
105 – **At the beginning of your relationship, was your partner/spouse:** **1.** Active, he/she was working **2.** Looking after the home or family; economically inactive ➔ ***FILTER*** **3.** Unemployed, searching for a job ➔ ***FILTER*** **4.** Pupil, student, apprentice ➔ ***FILTER*** **5.** Other inactive (ill, retired) ➔ ***FILTER***	|__|	|__|	|__|	|__|	|__|
106 – **Was he/she:** **1.** Higher-level occupation **2.** Skilled employee or worker **3.** Unskilled employee, worker, labourer **4.** Employer/ self-employed **5.** Helping family member	|__|	|__|	|__|	|__|	|__|
FILTER: ***Go to next relationship***					

- ***MORE RELATIONSHIPS*** ➔ ***Additional sheets;*** *Otherwise,* ***CHILDREN MODULE*** ➔ ***Next page***

MODULE: CHILDREN

200 – COUNT THE NUMBER OF CHILDREN IN GRID (COLUMN 2.2):
|__|__| *CHILDREN. DO NOT REPEAT THE QUESTION ABOUT THE NUMBER OF CHILDREN TO THE RESPONDENT.*
NOW I WOULD LIKE TO ASK A FEW QUESTIONS ABOUT YOUR CHILDREN. SO YOUR FIRST CHILD IS. . .

Questions	C 01	C 02	C 03	C 04	C 05
201B – 201D – Year of birth and possibly year of death *See GRID* *If child is alive, cross out year of death*	|__|__| |__|__| Birth Death	|__|__| |__|__| Birth Death	|__|__| |__|__| Birth Death	|__|__| |__|__| Birth Death	|__|__| |__|__| Birth Death
202 – No. of child *see GRID*	|__|__|	|__|__|	|__|__|	|__|__|	|__|__|
203 – Number of the relationship in which the child was born *see GRID – If birth occurred outside union; note 00*	|__|__|	|__|__|	|__|__|	|__|__|	|__|__|
First name of child *Information not retained in data entry*	. . . .	. . . .	. . . .	. . . .	. . . .
204 – Is this child a girl or a boy? **1.** Male **2.** Female	|__|	|__|	|__|	|__|	|__|
205 – In which district was he/she born? *Note answer in plain text*	. . . .	. . . .	. . . .	. . . .	. . . .
206 – What is/are his/her nationalities: Write down in plain text ALL nationalities held	. . . .	. . . .	. . . .	. . . .	. . . .
207 – Has this child always lived with you? **1.** Yes **→Go to next child** **2.** No	|__|	|__|	|__|	|__|	|__|
208 – How old was the child when they left home for the first time?	|__|__| years old	|__|__| years old	|__|__| years old	|__|__| years old	|__|__| years old
209 – Why did they leave? **1.** Looked after by someone else **2.** Studies **3.** Looking for work **4.** Apprenticeship/started work **5.** Marriage **6.** Other	|__|	|__|	|__|	|__|	|__|
210 – Where did they go? *Note name of district in plain text*	. . . .	. . . .	. . . .	. . . .	. . . .

• *FURTHER CHILDREN* → *Next page* *Otherwise,* ***HOUSING HISTORY*** → *Page 13*

Questions	C 06	C 07	C 08	C 09	C 10
201B – 201D – Year of birth and possibly year of death *See GRID* *If child is alive, cross out year of death*	|__|__| |__|__| Birth Death	|__|__| |__|__| Birth Death	|__|__| |__|__| Birth Death	|__|__| |__|__| Birth Death	|__|__| |__|__| Birth Death
202 – No. of child *see GRID*	|__|__|	|__|__|	|__|__|	|__|__|	|__|__|
203 – Number of the relationship in which the child was born *see GRID – If birth occurred outside union; note 00*	|__|__|	|__|__|	|__|__|	|__|__|	|__|__|
First name of child *Information not retained in data entry*	....	....	....	....	....
204 – Is this child a girl or a boy? **1.** Male **2.** Female	|__|	|__|	|__|	|__|	|__|
205 – In which district was he/she born? *Note answer in plain text*	....	....	....	....	....
206 – What is/are his/her nationalities: Write down in plain text ALL nationalities held	....	....	....	....	....
207 – Has this child always lived with you? **1.** Yes **→ Go to next child** **2.** No	|__|	|__|	|__|	|__|	|__|
208 – How old was the child when they left home for the first time?	|__|__| years old	|__|__| years old	|__|__| years old	|__|__| years old	|__|__| years old
209 – Why did they leave? **1.** Looked after by someone else **2.** Studies **3.** Looking for work **4.** Apprenticeship/started work **5.** Marriage **6.** Other	|__|	|__|	|__|	|__|	|__|
210 – Where did they go? *Note name of district in plain text*	....	....	....	....	....

- *FURTHER CHILDREN* → *Additional sheet* *Otherwise,* ***HOUSING HISTORY*** → *Next page*

MODULE: HOUSING HISTORY

LET'S COME BACK TO THE HOUSES WHERE YOU HAVE LIVED.
300 – *Count in GRID (column 3.1)*: **YOU HAVE LIVED IN |___|___| HOUSES.**
INTERVIEWER: In this module you have to fill in one column for each of the houses occupied by the respondent. Special case: If the respondent commutes between two different places of residence during a certain period of time: Fill in a column to describe each of the houses, place a curly bracket over the 2 columns and note the frequency of changeover (e.g. 9 months in dhaka; 3 months in khulna, or weekdays in boarding school; weekends at my Uncle's PLACE) . . .

Questions	**D 01**	**D 02**	**D 03**	**D 04**	**D 05**
301D – 301F – Years of arrival in and departure *see GRID* *If ongoing cross out the end year*	|__|__| |__|__| Start End	|__|__| |__|__| Start End	|__|__| |__|__| Start End	|__|__| |__|__| Start End	|__|__| |__|__| Start End
302 – Name of the DISTRICT *see GRID*	. . . .	. . . .	. . . .	. . . .	. . . .
303 – You lived then in "name of the TOWN or VILLAGE" *see GRID – in CAPITAL letters*	. . . .	. . . .	. . . .	. . . .	. . . .
304 – When you arrived in this house, were you: **1.** Living rent free **2.** Tenant (paying rent) **3.** Owner or leaseholder **4.** Resident in a hostel, student residence **5.** Other	|__|	|__|	|__|	|__|	|__|
305 – What type of housing was it? **1.** A room **2.** An apartment **3.** A traditional house **4.** A modern house (e.g. brick) **5.** Other, *Specify*	|__|	|__|	|__|	|__|	|__| . . . .

Questions	D 01	D 02	D 03	D 04	D 05
306 – *When you lived in this house* would you say that the financial situation of the household regarding the purchase of staple goods was. . . **1.** Always sufficient? **2.** Just sufficient? **3.** Often insufficient?					
307 – And relative to other people from your village/town, would you say that your living conditions were: **1.** Better? **2.** Equivalent? **3.** Less good?					
308 – *When you lived in this house*, did you experience any of the following: *Read* **:** **1.** Flooding **2.** Cyclone **3.** Riverbank/coastal erosion **4.** Drinking water shortage/pollution **5.** Salinisation **6.** Reduced crop yields **7.** Reduced fish catch **8.** Drought/lack of rain **9.** Erratic rainfall **FILTER: No: Leave blank and go to next residence period**					
309 – Did this influence your migration decision? **1.** Yes **2.** Partially **3.** No					

Questions	D 01	D 02	D 03	D 04	D 05
310 – *While you lived in this house* would you say that any of the following became more of a concern. . . *Read* : **1.** Flooding **2.** Cyclone **3.** Riverbank/coastal erosion **4.** Drinking water shortage/pollution **5.** Salinisation **6.** Reduced crop yields **7.** Reduced fish catch **8.** Drought/lack of rain **9.** Erratic rainfall **FILTER: No: Leave blank and go to next residence period**	|__| |__| |__| |__| |__| |__| |__|	|__| |__| |__| |__| |__| |__| |__|	|__| |__| |__| |__| |__| |__| |__|	|__| |__| |__| |__| |__| |__| |__|	|__| |__| |__| |__| |__| |__| |__|
311 – Did this influence your migration decision? **1.** Yes **2.** Partially **3.** No	|__|	|__|	|__|	|__|	|__|

- ***MORE HOUSES*** ➔ *Next page*
- *Otherwise*

312 **Of all of these places you have lived, which do you most consider your home?** |__|__|

313 **Is this where your father or mother lived?** **1.** Yes **2.** No |__|

313 **Do you consider it safe to live in this place?** **1.** Yes **2.** No |__|

314 **Do you want your children to live/grow up in this place?** **1.** Yes **2.** No |__|

- *Now go to **ACTIVITY AND EDUCATION HISTORY*** ➔ *Page **19***

Questions	D 06	D 07	D 08	D 09	D10
301D – 301F – Years of arrival in and departure *see GRID* *If ongoing cross out the end year*	|__|__| |__|__| Start End	|__|__| |__|__| Start End	|__|__| |__|__| Start End	|__|__| |__|__| Start End	|__|__| |__|__| Start End
302 – Name of the DISTRICT *see GRID*	....	....	....	....	....
303 – You lived then in "name of the TOWN or VILLAGE" *see GRID – in CAPITAL letters*	....	....	....	....	....
304 – When you arrived in this house, were you: **1.** Living rent free **2.** Tenant (paying rent) **3.** Owner or leaseholder **4.** Resident in a hostel, student residence **5.** Other	|__|	|__|	|__|	|__|	|__|
305 – What type of housing was it? **1.** A room **2.** An apartment **3.** A traditional house **4.** A modern house (e.g. brick) **5.** Other, *Specify*	|__|	|__|	|__|	|__|	|__| ....

Questions	D 06	D 07	D 08	D 09	D10
306 – *When you lived in this house* would you say that the financial situation of the household regarding the purchase of staple goods was. . . **1.** Always sufficient? **2.** Just sufficient? **3.** Often insufficient?	[ ]	[ ]	[ ]	[ ]	[ ]
307 – And relative to other people from your village/town, would you say that your living conditions were: **1.** Better? **2.** Equivalent? **3.** Less good?	[ ]	[ ]	[ ]	[ ]	[ ]
308 – *When you lived in this house*, did you experience any of the following: **1.** Flooding **2.** Cyclone **3.** Riverbank/coastal erosion **4.** Drinking water shortage/pollution **5.** Salinisation **6.** Reduced crop yields **7.** Reduced fish catch **8.** Drought/lack of rain **9.** Erratic rainfall **FILTER: No: Leave blank and go to next residence period**	[ ] [ ] [ ] [ ] [ ] [ ] [ ]	[ ] [ ] [ ] [ ] [ ] [ ] [ ]	[ ] [ ] [ ] [ ] [ ] [ ] [ ]	[ ] [ ] [ ] [ ] [ ] [ ] [ ]	[ ] [ ] [ ] [ ] [ ] [ ] [ ]
309 – Did this influence your migration decision? **1.** Yes **2.** Partially **3.** No	[ ]	[ ]	[ ]	[ ]	[ ]

Questions	D 06	D 07	D 08	D 09	D10
310 – *While you lived in this house* would you say that any of the following became more of a concern. . . **1.** Flooding **2.** Cyclone **3.** Riverbank/coastal erosion **4.** Drinking water shortage/pollution **5.** Salinisation **6.** Reduced crop yields **7.** Reduced fish catch **8.** Drought/lack of rain **9.** Erratic rainfall **FILTER: No: Leave blank and go to next residence period**	|__| |__| |__| |__| |__| |__| |__|	|__| |__| |__| |__| |__| |__| |__|	|__| |__| |__| |__| |__| |__| |__|	|__| |__| |__| |__| |__| |__| |__|	|__| |__| |__| |__| |__| |__| |__|
311 – Did this influence your migration decision? **1.** Yes **2.** Partially **3.** No	|__|	|__|	|__|	|__|	|__|

- ***MORE HOUSES*** ➔ *Additional sheets*
- *Otherwise*

312 Of all of these places you have lived, which do you most consider your home? |___|___|

313 Is this where your father or mother lived? **1.** Yes **2.** No |___|

313 Do you consider it safe to live in this place? **1.** Yes **2.** No |___|

314 Do you want your children to live/grow up in this place? **1.** Yes **2.** No |___|

- *Now go to* ***ACTIVITY AND EDUCATION HISTORY*** ➔ *Next Page*

ACTIVITY AND EDUCATION HISTORY

COLUMN 5

We will now talk about what you have been doing since your childhood: I would like to ask you about your periods of STUDY, of PROFESSIONAL TRAINING, of WORK, at HOME or if you were UNEMPLOYED, etc.

Note in the grid the primary activities lasting at least for 1 year (or equivalent duration to one academic year). Indicate, if there is enough space; the shorter activity periods in the column titled "Comments and Specifications" e.g. 5 months unemployed after dismissal.

- *1st activity/inactivity:* **What did you do at the age of 6? What was your primary activity?**
  GRID: Note the primary activity at 6 years: "school," "helped parents in the field," "at home," etc. . . .

  Until when did you continue (*adjust*) **going to school/staying at home/helping your parents. . . ?**
  GRID: Locate the year where the first change in the occupation occurs and draw an arrow to the 2nd occupation.
- *2nd activity/inactivity:* **And then; what did you do? And until when?**
  GRID: Note the new activity or inactivity at the line of the year when it begins and draw an arrow to the 3rd occupation.
- *Continue in the same way for each activity or inactivity period, up to the respondent's current situation.*

ATTENTION: *Always start a new activity period when Ego changes District, even if his or her activity remains the same.*

Study periods

- *Do not differentiate between different levels of schooling.*
- *. . . but consider University as a specific period*
- *Indicate possible interruptions in the education periods.*

Occupation periods: Consider as a change in the time period every change in activity consisting in:

- *A change in occupation, profession, status*
- *A change of employer*

Q16 – What is the highest level of qualification that you have obtained?

1. No schooling **2.** Junior school certificate **3.** Secondary school certificate **4.** Higher Secondary certificate **5.** Degree

MODULE – PERIODS OF ACTIVITY AND INACTIVITY

LET'S TALK IN SOME DETAIL ABOUT THE DIFFERENT EDUCATIONAL AND OCCUPATIONAL STATUSES YOU HAVE HAD IN YOUR LIFE. . .

400 – *Count (column 5) the different periods in GRID:*|___|___| *WITHOUT FORGETTING THE SCHOOLING AND ECONOMICALLY INACTIVE PERIODS.*

Questions	A 01	A 02	A 03	A 04	A 05
401S – 401E – Start and end years *see GRID* *If ongoing cross out the end year*	Start End	Start End	Start End	Start End	Start End
402 – During this period, you were *primarily:* 1. Active, you were working 2. Looking after the home or family; economically inactive → *406* 3. Unemployed, searching for a job → *406* 4. Pupil, student, apprentice → *406* 5. Other inactive (ill, retired) → *406*					
403 – What was your exact occupation during this period? What were your tasks? *Describe very precisely: occupation, level of qualification, sector*					
404 – Were you . . . 1. Higher-level occupation 2. Skilled employee or worker 3. Unskilled employee, worker, labourer 4. Employer/ self-employed 5. Helping family member, business or farm					

Questions	A 01	A 02	A 03	A 04	A 05
405 – All-in-all would you say that during this period you had enough to live on from day-to-day? **1.** Yes, absolutely **2.** It depended **3.** No, not at all	|__|	|__|	|__|	|__|	|__|
406 – At one moment or another during this period, did you receive . . . *Read:* **1.** A wage, income from your main activity? **2.** Income from moonlighting, small jobs, occasional employment? **3.** An unemployment benefit? **4.** A retirement pension, disability pension, other type of pension? **5.** Social benefits (family allowances, welfare benefits) **6.** A scholarship? **7.** Income from rents, interest or other capital income? **8.** Other resources? ***If no resource → Check off and go to the next period***	*No resource θ*	*No resource θ*	*No resource θ*	*No resource θ*	*No resource θ*
407 – What is the main reason you changed economic activity? *Write as clearly as possible the reason given*	. . . .	. . . .	. . . .	. . . .	. . . .

- *More periods of ACTIVITY or INACTIVITY* → *Next page*
- Otherwise, go to HISTORY OF ASSETS AND BUSINESSES → *Page* **24**

Questions	A 06	A 07	A 08	A 09	A 10
401S – 401E – Start and end years *see GRID* *If ongoing cross out the end year*	Start End	Start End	Start End	Start End	Start End
402 – During this period, you were *primarily:* **1.** Active, you were working **2.** Looking after the home or family; economically inactive ➔ **406** **3.** Unemployed, searching for a job ➔ **406** **4.** Pupil, student, apprentice ➔ **406** **5.** Other inactive (ill, retired) ➔ **406**					
403 – What was your exact occupation during this period? What were your tasks? *Describe very precisely: occupation, level of qualificction, sector*					
404 – Were you . . . **1.** Higher-level occupation **2.** Skilled employee or worker **3.** Unskilled employee, worker, labourer **4.** Employer/ self-employed **5.** Helping family member, business or farm					

Questions	A 06	A 07	A 08	A 09	A 10
405 – All-in-all would you say that during this period you had enough to live on from day-to-day? **1.** Yes, absolutely **2.** It depended **3.** No, not at all	|__|	|__|	|__|	|__|	|__|
406 – At one moment or another during this period, did you receive . . . ***Read:*** **1.** A wage, income from your main activity? **2.** Income from moonlighting, small jobs, occasional employment? **3.** An unemployment benefit? **4.** A retirement pension, disability pension, other type of pension? **5.** Social benefits (family allowances, welfare benefits) **6.** A scholarship? **7.** Income from rents, interest or other capital income? **8.** Other resources? ***If no resource ➔ Check off and go to the next period***	|__||__| |__||__| |__||__| |__||__| *No resource θ*	|__||__| |__||__| |__||__| |__||__| *No resource θ*	|__||__| |__||__| |__||__| |__||__| *No resource θ*	|__||__| |__||__| |__||__| |__||__| *No resource θ*	|__||__| |__||__| |__||__| |__||__| *No resource θ*
407 – What is the main reason you changed economic activity? *Write as clearly as possible the reason given*	. . . . |__|__|__|	. . . . |__|__|__|	. . . . |__|__|__|	. . . . |__|__|__|	. . . . |__|__|__|

- *More periods of ACTIVITY or INACTIVITY* ➔ *Additional sheets*
- Otherwise, go to HISTORY OF ASSETS AND BUSINESSES ➔ *Next page*

MODULE: HISTORY OF ASSETS AND BUSINESSES OWNED

Now we will talk about the assets or businesses that you may have bought over your lifetime, or that you may have received or inherited from somebody.

1. Are you CURRENTLY owner . . .			**2. And in the past, have you been owner, here or elsewhere. . .**			**Total**
. . . of one or several plots of land (agricultural land, building plot, or under construction)	**1.** Yes → How many? **2.** No → *Note 00*	Q17PC |__|__|	. . . of plots that you don't own anymore?	**1.** Yes → How many? **2.** No → *Note 00*	Q17PP |__|__|	
. . . of one or several housing units (house, apartment . . .)?	**1.** Yes → How many? **2.** No → *Note 00*	Q17DC |__|__|	. . . of housing units that you don't own anymore?	**1.** Yes → How many? **2.** No → *Note 00*	Q17DP |__|__|	
. . . of a business, venture, commercial premises even on a rental basis (shop, workshop, taxis, rickshaw, boat . . .)?	**1.** Yes → How many? **2.** No → *Note 00*	Q17BC |__|__|	. . . of a business, a venture, commercial premises even on a rental basis that you don't own anymore?	**1.** Yes → How many? **2.** No → *Note 00*	Q17BP |__|__|	
Total		Q17TC |__|__|			Q17TP |__|__|	Q17TOT |__|__|

FILTR: *If NO ASSET (Q17TOT = 0) → Go to TRANSFERS, Page 27*
Otherwise → Fill out one column per owned asset

Follow the order of the table: plots of land, then housing units, then businesses currently owned before continuing with the assets owned in the past.

Questions	AS01	AS 02	AS 03	AS 04	AS 05
LET'S FIRST TALK ABOUT YOUR. . . *Encircle the type of asset* **501 – How did you obtain this asset?** **1.** You bought it? **2.** You inherited it? **3.** You were given it? **4.** You reclaimed it? **5.** For another reason? Specify	Plot – House – Business |__|	Plot – House – Business |__|	Plot – House – Business |__|	Plot – House – Business |__|	Plot – House – Business |__| . . . .
501S – When (in which year) did you obtain it?	|__|__| Start	|__|__| Start	|__|__| Start	|__|__| Start	|__|__| Start
FILTER: If Ego is no longer owner of the asset → 501E ***Otherwise → 503***					
501E – And until when did you own this asset?	|__|__| End	|__|__| End	|__|__| End	|__|__| End	|__|__| End
502 – You don't own this asset anymore because . . . **1.** You sold it? **2.** You have donated/bequeathed it? **3.** You went bankrupt? **4.** For another reason? Specify	|__|	|__|	|__|	|__|	|__| . . . .
503 – Is the asset we are talking about: ***Plot of land*** **1.** A building plot, or with a building currently under construction **→ 508** **2.** A plot for agricultural use **→ 504** ***Housing unit*** **3.** A traditional house **→ 505** **4.** A modern house or apartment **→ 505** **5.** An apartment block **→ 505** ***Businesses and ventures*** **6.** A business, commercial premises (shop, workshop . . .) **→ 505** **7.** A business, venture without walls (patent, goodwill & tools & merchandise, taxis, rickshaws, boats . . .) **→ 505**	|__|__|	|__|__|	|__|__|	|__|__|	|__|__|

Questions	AS01	AS 02	AS 03	AS 04	AS 05
504 – Most of the time, this plot has been used . . . **1.** As grazing land/pasture **➔ 509** **2.** As an orchard **➔ 509** **3.** For market gardening **➔ 509** **4.** For irrigated crop-growing **➔ 509** **5.** For other types of crop **➔ 509** **6.** For aquaculture **➔ 509** **7.** Has been unused **➔ 510**	|__|	|__|	|__|	|__|	|__|
505 – Most of the time, this asset has been: **1.** Rented out (dwelling, commercial premises)?**➔ 510** **2.** Operated (business . . .)? **➔ 506** **3.** Used free of charge for personal use? **➔ 509** **4.** Unoccupied, unused? **➔ 510**	|__|	|__|	|__|	|__|	|__|
506 – What is/was the activity performed? **Note response in plain text**	. . . .	. . . .	. . . .	. . . .	. . . .
507 – This asset has been operated or used . . . *Read out* **1.** By yourself? **2.** By family members? **3.** By other persons?		__
__
__			__
__
__			__
__
__			__
__
__			__
__
__	
508 – In which District is this asset located? *Note answer in plain text*	. . . .	. . . .	. . . .	. . . .	. . . .

- ***MORE ASSETS*** ➔ *Additional sheets*
- *Otherwise, Go to TRANSFERS* ➔ *Next page*

TRANSFERS **COLUMNS 6 AND 7**

Q18 - Have there been periods at any time of your life during which you used to regularly send money to somebody who was living in a different place (e.g. District, or abroad) from the one where you were at the time?

1. Yes → **from which year(s) to which year(s)?**

→ **And in what District did the persons to whom you sent money live?**
GRID: Note "TRO + Name of the District" at the start year in column 6 and draw an arrow to the end of this period.

→ **Have there been further periods when you used to send money regularly?**

2. No

Q19 – Have there been any periods at any time of your life during which you used to regularly *receive* money from somebody who was living in a different place (i.e. District, or abroad) from the one you were living in at the time?

1. Yes → **from which year(s) to which year(s)?**

→ **And in what District did the persons from whom you received money live?**
GRID: Note "TRI + Name of the District" at the start year in column 7 and draw an arrow to the end of this period.

→ **Have there been further periods when you used to receive money regularly?**

2. No

STAYS OF LESS THAN A YEAR OUTSIDE YOUR TOWN/VILLAGE **COLUMN 8**

Q20 – Have there been periods where you have stayed for periods of less than a year, but more than a month outside the District you were/are living in?

1. Yes
2. No → { ***- If Ego has already lived outside District (column 3.2)*** → ***Module LONG AND SHORT STAYS OUTSIDE DISTRICT, page 29***
- If ego has never lived outside District → ***END of the interview; Note the time on page 33***

In which Districts have you stayed? *List the Districts*

1.	**5.**	**9.**	**13.** ....
2.	**6.**	**10.**	**14.** ....
3.	**7.**	**11.**	**15.** ....
4.	**8.**	**12.**	**16.** ....

- *1st District:* **In which year did you go there for the first time?**
  GRID: Note the name of the District in the appropriate year
  Did you visit this country again later on, staying again for more than a month, but less than a year?
  If yes: **In which year(s)?**
  In GRID: Note ALL stays of more than a month but less than a year in this District in the appropriate year(s)
- *2nd District: CONTINUE IN THIS WAY for each District.*

ATTENTION: - If during a period of several years the respondent visits a District or several Districts for the same reason every year: GROUP these stays. Note the District or Districts at the beginning of the time period and draw an arrow to the end of the period.

- *DON'T FORGET to explore other possible stays outside this time period.*

MODULE: LONG AND SHORT STAYS OUTSIDE DISTRICT OF BIRTH

(= SHORT STAYS + STAYS OF MORE THAN 1 YEAR)

LET'S LOOK IN SOME MORE DETAIL AT THE HISTORY OF YOUR STAYS OUTSIDE YOUR DISTRICT OF BIRTH.

600S – Count (column 8) every short stay *(recount a District if it is cited several times in GRID but the trips are not grouped):* |___|___|

600L – Count (column 3.2) every STAY OF MORE THAN 1 YEAR *(count a District several times if Ego went there repeatedly):* |___|___|

600TOT – TOTAL : |___|___|

Fill in one column per stay, category by category (SHORT, then LONG), following a chronological order within each category.

Questions	S01	S02	S03	S04	S05
601S – 601E – Start and end years of the stay – see *GRID* *If stay is ongoing, cross out end date*	|__|__||__|__| Start End	__|__||__|__| Start End	|__|__||__|__| Start End	|__|__||__|__| Start End	|__|__||__|__| Start End
602 – DISTRICT OF STAY/ARRIVAL **see** *GRID, columns 3.2 and 8*	....	....	....	....	....
603 – In which District were you just before arriving in "District of stay"? *Do not rely on GRID*	....	....	....	....	....
604 – For what reasons did you leave this District? ***Note precisely and verbatim the entire response***	....	....	....	....	....

Questions	S01	S02	S03	S04	S05
605 – And for what reasons did you choose to go to "District of stay" ***rather than to anywhere else?*** ***Note precisely and verbatim the entire response***	. . . .	. . . .	. . . .	. . . .	. . . .
606 – Did you need a permit or other documentation to live or work in "District of stay" . . . **1.** A residence permit **2.** A permit to work **3.** You didn't need any permission ➔ **608**	|__||__|	|__||__|	|__||__|	|__||__|	|__||__| . . . .
607 – **Did you have this permit or documentation?** **1.** Yes **2.** For some of the time **3.** No	|__||__||__| |__||__||__|	|__||__||__| |__||__||__|	|__||__||__| |__||__||__|	|__||__||__| |__||__||__|	|__||__||__| |__||__||__| . . . .
608 – Did you travel. . . **Read:** **1.** With your father or mother? **2.** With your partner(s)? **3.** With another family member? **4.** With one or several friends? **5.** With the whole community? **6.** With somebody else? *Specify* **0.** Alone?	|__||__| |__||__| |__||__|	|__||__| |__||__| |__||__|	|__||__| |__||__| |__||__|	|__||__| |__||__| |__||__|	|__||__| |__||__| |__||__| . . . .

Questions	S01	S02	S03	S04	S05
609 – Who decided about your trip/migration? **Read:** **0.** Yourself **1.** Your father or mother? **2.** Your partner? **3.** Your in-laws? **4.** Another family member? **5.** Your employer? **6.** The community? **7.** Somebody else? *Specify*					
610 – And who helped to finance your migration? **Read:** **0.** Yourself **1.** Your father or mother? **2.** Your partner? **3.** Your in-laws? **4.** Another family member? **5.** Your employer? **6.** The community **7.** Somebody else? *Specify*					
611 – In the year you moved, had any of the following happened in the place you were living (before you moved)? **Read:** **1.** Flooding **2.** Cyclone **3.** Riverbank/coastal erosion **4.** Drinking water shortage/pollution **5.** Salinisation **6.** Reduced crop yields **7.** Reduced fish catch **8.** Drought/lack of rain **9.** Erratic rainfall					

Questions	S01	S02	S03	S04	S05
612 – While you lived in "District of stay," did you make any monetary or in-kind donations to help the inhabitants of your home village, e.g. to build facilities? **1.** Yes **2.** No ➔ *Go to next stay*	|__|	|__|	|__|	|__|	|__|
613 – Did you contribute to building. . . **1.** A school? **2.** A health centre? **3.** A borehole (to supply water)? **4.** Flood defences? **5.** A drainage system? **6.** An irrigation system? **7.** A mosque or religious building? **8.** A cyclone shelter? **9.** Something else? *Specify*	|__| |__| |__| |__| |__| |__| |__|	|__| |__| |__| |__| |__| |__| |__|	|__| |__| |__| |__| |__| |__| |__|	|__| |__| |__| |__| |__| |__| |__|	|__| |__| |__| |__| |__| |__| |__| ....

- ***MORE STAYS*** ➔ *Additional sheets*
- *Otherwise. . .*

We are now at the end of the interview. I thank you very much for your participation.

Would you like to make any comments or give us your opinion about this questionnaire or this study?

....

....

END TIME : |___|___| hr |___|___| min

INTERVIEWER'S OBSERVATIONS – TO BE FILLED OUT AFTER THE INTERVIEW

E1 – This person was:

1. Easily persuaded to participate ➔ **E2**
2. A bit difficult to persuade
3. Very difficult to persuade

E1A – For what reasons was he/she reluctant to participate? And which arguments enabled you to convince him/her in the end?

. . . .

E2 – The reception by the respondent was:

1. Very good over the entire interview
2. Good, but reluctant on certain questions ➔ **E2R – Which ones?** . . . .
3. Quite reluctant or suspicious over the entire duration of the interview
4. Other: *Specify*: . . . .

E4 – And were any other person(s) present during the interview?

1. yes
2. no ➔ E5

E4I – Did you have the impression that this presence influenced the respondent in his/her answers?

1. Yes, the entire questionnaire
2. Yes, certain parts of the questionnaire ➔ **E4Q – Which ones (specify the nb of the questions)?**

. . . .

3. No

E6 – According to you, was the respondent's general comprehension of the questions:

1. Very good
2. Adequate, but not perfect
3. Bad

E7 – And did the respondent have problems answering certain questions?

1. Yes ➔ **E7D – Which ones? No**: . . . .

2. No

INFORMATION TO TRANSCRIBE AFTER THE INTERVIEW BASED ON THE BIOGRAPHICAL GRID

MODULE: MIGRATIONS AMONG FAMILY OR CONTACT CIRCLE MEMBERS SEE GRID COLUMN S4

900 – Count in GRID the number of family or contact circle members who have lived outside District of Ego's birth |___|___| and fill in one column per person. (in principle, number equal to Q15TOT).

MIGRANTS IN THE FAMILY/NETWORK	M1	M2	M3	M4	M5
901 – Relationship: the person is Ego's *Code:* **1.** Partner + No. **2.** Son/daughter + No. **3.** Father/mother **4.** Brother/sister **5.** Other relative, *Specify* **6.** Friend **7.** Other, *Specify* *If the person is Ego's child or partner indicate his or number given in GRID.*	Relationship: |___|___| . . . . No. Partner (Q101): |___|___| No. Child (Q202): |___|___|	Relationship: |___|___| . . . . No. Partner (Q101): |___|___| No. Child (Q202): |___|___|	Relationship: |___|___| . . . . No. Partner (Q101): |___|___| No. Child (Q202): |___|___|	Relationship: |___|___| . . . . No. Partner (Q101): |___|___| No. Child (Q202): |___|___|	Relationship: |___|___| . . . . No. Partner (Q101): |___|___| No. Child (Q202): |___|___|
902 – Sex: 1. Male 2. Female	|___|	|___|	|___|	|___|	|___|
903M – Year in which they met *Cross out if the person is not a partner or a friend*	|___|___|	|___|___|	|___|___|	|___|___|	|___|___|
903D – Year of death *Cross out if the person is not deceased*	|___|___|	|___|___|	|___|___|	|___|___|	|___|___|
904 – District 1 (1st District outside of place of birth). *In plain text and CAPITALS*	. . . .	. . . .	. . . .	. . . .	. . . .
904S – 904E – Start and end year *Cross out if still there*	|___|___||___|___| Start End	|___|___||___|___| Start End	|___|___||___|___| Start End	|___|___||___|___| Start End	|___|___||___|___| Start End
905 – District 2 (2nd District outside of place of birth). *In plain text and CAPITALS*	. . . .	. . . .	. . . .	. . . .	. . . .
905S – 905E – Start and end year *Cross out if still there*	|___|___||___|___| Start End	|___|___||___|___| Start End	|___|___||___|___| Start End	|___|___||___|___| Start End	|___|___||___|___| Start End
906 – District 3 (3rd District outside of place of birth) *In plain text and CAPITALS*	. . . .	. . . .	. . . .	. . . .	. . . .
906S – 906E – Start and end year *Cross out if still there*	|___|___||___|___| Start End	|___|___||___|___| Start End	|___|___||___|___| Start End	|___|___||___|___| Start End	|___|___||___|___| Start End
907 – District 4 (4th District outside of place of birth) *In plain text and CAPITALS*	. . . .	. . . .	. . . .	. . . .	. . . .
907S – 907E – Start and end year *Cross out if still there*	|___|___||___|___| Start End	|___|___||___|___| Start End	|___|___||___|___| Start End	|___|___||___|___| Start End	|___|___||___|___| Start End

• ***More migrants*** → *next page* • ***MODULE CITIZENSHIP*** → *Page 38*

MIGRANTS IN THE FAMILY/NETWORK	M6	M7	M8	M9	M10
901 – Relationship: the person is Ego's *Code:* **1.** Partner + No. **2.** Son/daughter + No. **3.** Father/mother **4.** Brother/sister **5.** Other relative, *Specify* **6.** Friend **7.** Other, *Specify* *If the person is Ego's child or partner indicate his or number given in GRID.*	Relationship: ___ ___ .... No. Partner (Q101): ___ ___ No. Child (Q202): ___ ___	Relationship: ___ ___ .... No. Partner (Q101): ___ ___ No. Child (Q202): ___ ___	Relationship: ___ ___ .... No. Partner (Q101): ___ ___ No. Child (Q202): ___ ___	Relationship: ___ ___ .... No. Partner (Q101): ___ ___ No. Child (Q202): ___ ___	Relationship: ___ ___ .... No. Partner (Q101): ___ ___ No. Child (Q202): ___ ___
902 – Sex: 1. Male 2. Female	___	___	___	___	___
903M – Year in which they met *Cross out if the person is not a partner or a friend*	___ ___	___ ___	___ ___	___ ___	___ ___
903D – Year of death *Cross out if the person is not deceased*	___ ___	___ ___	___ ___	___ ___	___ ___
904 – District 1 (1st District outside of place of birth). *In plain text and CAPITALS*	....	....	....	....	....
904S – 904E – Start and end year *Cross out if still there*	___ ___ ___ ___ Start End	___ ___ ___ ___ Start End	___ ___ ___ ___ Start End	___ ___ ___ ___ Start End	___ ___ ___ ___ Start End
905 – District 2 (2nd District outside of place of birth). *In plain text and CAPITALS*	....	....	....	....	....
905S – 905E – Start and end year *Cross out if still there*	___ ___ ___ ___ Start End	___ ___ ___ ___ Start End	___ ___ ___ ___ Start End	___ ___ ___ ___ Start End	___ ___ ___ ___ Start End
906 – District 3 (3rd District outside of place of birth) *In plain text and CAPITALS*	....	....	....	....	....
906S – 906E – Start and end year *Cross out if still there*	___ ___ ___ ___ Start End	___ ___ ___ ___ Start End	___ ___ ___ ___ Start End	___ ___ ___ ___ Start End	___ ___ ___ ___ Start End
907 – District 4 (4th District outside of place of birth) *In plain text and CAPITALS*	....	....	....	....	....
907S – 907E – Start and end year *Cross out if still there*	___ ___ ___ ___ Start End	___ ___ ___ ___ Start End	___ ___ ___ ___ Start End	___ ___ ___ ___ Start End	___ ___ ___ ___ Start End

• ***More migrants*** ➔ *next page* • ***MODULE CITIZENSHIP*** ➔ *Page 38*

MIGRANTS IN THE FAMILY/NETWORK	M11	M12	M13	M14	M15
901 – Relationship: the person is Ego's *Code:* **1.** Partner + No. **2.** Son/daughter + No. **3.** Father/mother **4.** Brother/sister **5.** Other relative, *Specify* **6.** Friend **7.** Other, *Specify* *If the person is Ego's child or partner indicate his or number given in GRID.*	Relationship: ___ ___ . . . . No. Partner (Q101): ___ ___ No. Child (Q202): ___ ___	Relationship: ___ ___ . . . . No. Partner (Q101): ___ ___ No. Child (Q202): ___ ___	Relationship: ___ ___ . . . . No. Partner (Q101): ___ ___ No. Child (Q202): ___ ___	Relationship: ___ ___ . . . . No. Partner (Q101): ___ ___ No. Child (Q202): ___ ___	Relationship: ___ ___ . . . . No. Partner (Q101): ___ ___ No. Child (Q202): ___ ___
902 – Sex: 1. Male 2. Female	___	___	___	___	___
903M – Year in which they met *Cross out if the person is not a partner or a friend*	___ ___	___ ___	___ ___	___ ___	___ ___
903D – Year of death *Cross out if the person is not deceased*	___ ___	___ ___	___ ___	___ ___	___ ___
904 – District 1 (1st District outside of place of birth). *In plain text and CAPITALS*	. . . .	. . . .	. . . .	. . . .	. . . .
904S – 904E – Start and end year *Cross out if still there*	___ ___ ___ ___ Start End	___ ___ ___ ___ Start End	___ ___ ___ ___ Start End	___ ___ ___ ___ Start End	___ ___ ___ ___ Start End
905 – District 2 (2nd District outside of place of birth). *In plain text and CAPITALS*	. . . .	. . . .	. . . .	. . . .	. . . .
905S – 905E – Start and end year *Cross out if still there*	___ ___ ___ ___ Start End	___ ___ ___ ___ Start End	___ ___ ___ ___ Start End	___ ___ ___ ___ Start End	___ ___ ___ ___ Start End
906 – District 3 (3rd District outside of place of birth) *In plain text and CAPITALS*	. . . .	. . . .	. . . .	. . . .	. . . .
906S – 906E – Start and end year *Cross out if still there*	___ ___ ___ ___ Start End	___ ___ ___ ___ Start End	___ ___ ___ ___ Start End	___ ___ ___ ___ Start End	___ ___ ___ ___ Start End
907 – District 4 (4th District outside of place of birth) *In plain text and CAPITALS*	. . . .	. . . .	. . . .	. . . .	. . . .
907S – 907E – Start and end year *Cross out if still there*	___ ___ ___ ___ Start End	___ ___ ___ ___ Start End	___ ___ ___ ___ Start End	___ ___ ___ ___ Start End	___ ___ ___ ___ Start End

- ***More migrants*** ➔ *additional sheets* • ***MODULE CITIZENSHIP*** ➔ *Next page*

MODULE CITIZENSHIP

SEE GRID COLUMN 9

1200 – Count in GRID the number of periods during which ego had one or several nationalities: |___|___|

	Start and end years	**Nationalities held** *in CAPITAL letters*
1200S – 1200F – Nationality or nationalities by birth	|__|__| |__|__| Start End	N: .... N: .... N: .... N: ....
1201S – 1201F – 1st change *Cross out if no change*	|__|__| |__|__| Start End	N: .... N: .... N: .... N: ....
1202S – 1202F – 2nd change *Cross out if no change*	|__|__| |__|__| Start End	N: .... N: .... N: .... N: ....
1203S – 1203F – 3rd change *Cross out if no change*	|__|__| |__|__| Start End	N: .... N: .... N: .... N: ....

MODULE TRANSFERS SENT

SEE GRID COLUMN 6

1600 – Count in GRID the periods of REGULAR TRANSFERS SENT: |___|___| and fill in one column for each period

	1st period TRO	**2nd period TRO**	**3rd period TRO**	**4th period TRO**	**5th period TRO**
1601S – 1601E – Start and end years of transfers *Cross out if ongoing*	|__|__| |__|__| Start End	|__|__| |__|__| Start End	|__|__| |__|__| Start End	|__|__| |__|__| Start End	|__|__| |__|__| Start End
1601P – Destination countries of transfers *Note in plain text all countries*	....	....	....	....	....

MODULE TRANSFERS RECEIVED

SEE GRID COLUMN 7

1700 – Count in GRID the periods of REGULAR TRANSFERS RECEIVED: |___|___| and fill in one column for each period

	1st period TRI	**2nd period TRI**	**3rd period TRI**	**4th period TRI**	**5th period TRI**
1701S – 1701E – Start and end years of transfers *Cross out if ongoing*	|__|__|__|__| Start End	|__|__|__|__| Start End	|__|__|__|__| Start End	|__|__|__|__| Start End	|__|__|__|__| Start End
1701P – Destination countries of transfers *Note in plain text all countries*	....	....	....	....	....

END – THANK YOU.

References

Abrar, C. R., and Azad, S. N. (2004). *Coping with displacement: Riverbank erosion in northwest Bangladesh.* Dhaka: RDRS, North Bengal Institute, NBI and Refugee and Migratory Movements Research Unit.

Abu, M., Codjoe, S. N. A., and Sward, J. (2014). Climate change and internal migration intentions in the forest-savannah transition zone of Ghana. *Population and Environment* 35: 341–364.

Adams, N., Dasgupta, S., and Sarraf, M. (2011a). *The cost of adapting to extreme weather events in a changing climate*, Bangladesh development series paper; no. 28. Washington, DC: The Worldbank. [online] Available at: http://documents.worldbank.org/curated/en/2011/12/16203403/cost-adapting-extreme-weather-events-changing-climate [Accessed on 17 July 2015.]

Adams, N., Dasgupta, S., and Sarraf, M. (2011b). *The cost of adapting to extreme weather events in a changing climate*, Bangladesh development series paper; no. 28. Annexes, Washington, DC: The Worldbank.

Adger, W. N., Arnell, N. W., and Tompkins, E. L. (2005). Successful adaptation to climate change across scales. *Global Environmental Change* 15: 77–86.

Afsar, R. (2003). *Internal migration and the development nexus: The case of Bangladesh.* Paper presented at the Regional conference on migration, development and pro-poor policy choices in Asia, organized by the Refugee and Migratory Movements Research Unit and DfID, UK at Dhaka, Bangladesh, 22–24 June.

Agrawala, S., Ota,T., Ahmed, A.U., Smith, S. and van Aalst (2003). Development and climate change in Bangaldesh: focus on coastal flooding and the Sundarbans, Paris: OECD

Ahmad, A. U., Hassan, S. R., Etzold, B., and Neelormi, S. (2012). *Where the rain falls project – Case study: Bangladesh*, Results from Kurigram district, Rangpur division, Report No. 2. Bonn: UNU-EHS.

Ahmad, N. (2012). *Gender and climate change in Bangladesh: The role of institutions in reducing gender gaps in adaptation program.* Social development working papers and a summary of ESW report no. P125705, Paper No. 126/March. Washington DC: The World Bank.

Ajzen, I. (1988). *Attitudes, Personality, and Behaviour*. Milton Keynes: Open University Press.

———. (1991). The theory of planned behavior. *Organizational Behavior and Human Decision Processes* 50: 179–211.

Ajzen, I., and Fishbein, M. (1980). *Understanding attitudes and predicting social behavior*. Englewood Cliffs, NJ: Prentice Hall.

Alam, E., and Collins, A. E. (2010). Cyclone disaster vulnerability and response experiences in coastal Bangladesh. Disasters 34: 931–954.
Allison, P. D. (1982). Discrete-time methods for the analysis of event histories. Sociological Methodology 13: 61–98.
Amin, S., Diamond, I., Naved, R. T., & Newby, M. (1998). Transition to adulthood of female garment-factory workers in Bangladesh. Studies in Family Planning, 29(2), 185–200.
Asian Development Bank (2011). Policy dialogues on climate induced migration. Geneva (9 June 2011) and Bangkok (16–17 June 2011), Discussion Paper. [online] Available at: www.iom.int/jahia/webdav/shared/shared/mainsite/activities/env_degradation/Discussion-Paper-Policy-Dialogues.pdf [Accessed on 23 November 2015.]
———. (2012). Addressing climate change and migration in Asia and the Pacific. Mandaluyong City: ADB.
Associated Press (2012). India pledges zero casualties on Bangladesh border. Dawn.com, World, Urdu edition, Associated Press, 29 September. [online] Available at: http://dawn.com/2012/09/29/india-pledges-zero-casualties-on-bangladesh-border/ [Accessed on 27 October 2012.]
Balá ž, V., and Williams, A. M. (2007). Path-dependency and path-creation: Perspectives on migration trajectories: The economic experiences of Vietnamese migrants in Slovakia. International Migration 45: 37–66.
Banerjee, L. (2007). Effect of flood on agricultural wages in Bangladesh: An empirical analysis. World Development 35(11): 1989–2009.
Banerjee, S., Black, R., and Kniveton, D. (2012). Migration as an effective mode of adaptation to cli-mate change. Policy paper for the European Commission. Brighton: Sussex Centre for Migration Research, University of Sussex.
Bangladesh Bank (2015). Wage earners remittance inflows: Country wise (Monthly). Available at: www.bb.org.bd/econdata/wagermidtl.php [Accessed on March 31, 2016]
Bangladesh Bureau of Statistics (2011). Population & housing census 2011: Preliminary results, Bangladesh Bureau of Statistics, Statistics Division. Dhaka: Ministry of Planning, Government of the People's Republic of Bangladesh.
———. (2012). Population & housing census 2011: National report, Volume 4, Socio-economic and demographic report, December 2012, Statistics and Informatics Division. Dhaka: Ministry of Planning, Government of the People's Republic of Bangladesh.
———. (2014). Population & housing census 2011: National volume – 3, Urban area report, August 2014, Statistics and Informatics Division. Dhaka: Ministry of Planning, Government of the People's Republic of Bangladesh.
Barnett, J., and Webber, M. (2010). Accommodating migration to promote adaptation to climate change, World Bank Policy Research Working Paper no. WPS 5270, April. Washington, DC: World Bank.
Barrios, S., Bertinelli, L., and Strobl, E. (2006). Climatic change and rural – urban migration: The case of sub-Saharan Africa. Journal of Urban Economics 60: 357–371.
Beratan, K. K. (2007). A cognition-based view of decision processes in complex social – ecological systems. Ecology and Society 12: 27.
Biermann, F., and Boas, I. (2010). Preparing for a warmer world. Towards a global governance system to protect climate refugees. Global Environmental Politics 10: 60–88.
Bilsborrow, R. E. (2009). Collecting data on the migration-environment nexus. In F. Laczko and C. Aghazarm (eds.), Migration, the environment and climate change: Assessing the evidence, Migration research series paper. Geneva: International Organization for Migration: 115–196.

Bilsborrow, R. E., and Okoth-Ogendo, H. W. O. (1992). Population-driven changes in land-use in developing countries. Ambio 21: 37–45.

Biswas, S., and Chowdhury, M. A. A. (2012). Climate change induced displacement and migration in Bangladesh: The need for rights-based solutions. Refugee Watch 39 & 40: 157–180.

Black, R. (1998). Refugees, environment and development. London: Longman.

———. (2001). Environmental refugees: Myth or reality? New issues in refugee research, Working paper 34. Geneva: UNHCR.

Black, R., Adger, N., Arnell, N. W., Dercon, S., Geddes, A., and Thomas, D. (2011a). The effect of environmental change on human migration. Global Environmental Change 21: S3–S11.

Black, R., Arnell, N. W., Adger, N. W., Thomas, D., and Geddes, A. (2013). Migration, immobility and displacement outcomes following extreme events. Environmental Science & Policy 27s: s32–s43.

Black, R. and Collyer, M. (2014) 'Populations 'trapped' at times of crisis. Forced Migration Review 45: 52 – 56

Black, R., Kniveton, D., and Schmidt-Verkerk, K. (2011b). Migration and climate change: Towards an integrated assessment of sensitivity. Environment and Planning A 43: 431–450.

Blackburn, Simon (2008). Oxford Dictionary of Philosophy, second edition revised. Oxford: Oxford University Press

Boano, C., Zetter, R., and Morris, T. (2007). Environmentally displaced people: Understanding the linkages between environmental change, livelihoods and forced migration, Forced Migration Policy Briefing 1. Refugee Studies Centre. Oxford: Department of International Development, University of Oxford.

Bohra-Mishra, P., Oppenheimer, M., and Hsiang, S. M. (2014). Nonlinear permanent migration response to climatic variations but minimal response to disasters. Proceedings of the National Academy of Sciences 111: 9780–9785.

Bourdieu, P. (1991). Language and symbolic power. Cambridge, MA: Harvard University Press.

Brown, L., Mcgrath, P., and Stokes, B., (1976). *Twenty two dimensions of the population problem*, Worldwatch Paper 5, Washington DC: Worldwatch Institute

Brown, O (2008) Migration and Climate Change, Geneva: IOM

Cannon, T., and Mueller-Mahn, D. (2010). Vulnerability, resilience and development discourses in context of climate change. Natural Hazards 55: 621–635.

Cannon, T., Schipper, L., Bankoff, G., and Krüger, F. (eds.) (2014). World disasters report 2014: Focus on culture and risk. Geneva: IFRC.

CARE-Bangladesh and DFID (2002). The findings of the Northwest Ruala livelihoods baseline – 2002 livelihood monitoring project. Dhaka, Bangladesh, cited in Hossain, M. I., Khan, I. A., and Seeley, J. (2003). Surviving on their feet: Charting the mobile livelihoods of the poor in rural Bangladesh. Paper presented in the conference, Staying poor: Chronic poverty and development policy, University of Manchester, April.

Carr, D. L., Lopez, A. C., and Bilsborrow, R. E. (2009). The population, agriculture, and environment nexus in Latin America: Country-level evidence from the latter half of the twentieth century. Population and Environment 30: 222–246.

Castles, S. (2002). Environmental change and forced migration: Making sense of the debate, New issues in refugee research, Working paper No. 70. Geneva: UNHCR.

Castles, S., De Haas, H., Miller, M. (2014). The Age of Migration: International Population Movements in the Modern World (Fifth Edition). New York: Guilford Press

Chowdhury, S., Mobarak, A. M., and Bryan, G. (2009). Migrating away from a seasonal famine: A randomized intervention in Bangladesh, Human development research paper 2009/41. New York: UNDP.
Christian Aid. (2009). Human tide: The real migration crisis. London: Christian Aid.
Climate and Development Knowledge Network (2011). Bangladesh's comprehensive disaster management programme, Inside stories on climate compatible development. [online] Available at: http://cdkn.org/resource/cdkn-inside-story-bangladesh's-comprehensive-disaster-management-programme/ [Accessed on 9 March 2015.]
Collins, A. E. (2013). Applications of the disaster risk reduction approach to migration influenced by environmental change. Environmental Science & Policy 27: S112–S125.
Cooke, E., Martin, M., and Kniveton, D. (2013). Migration as a response to climatic stresses and shocks in Bangladesh: A quantitative study, Working paper 3. Dhaka: Refugee and Migratory Movements Research Unit, and Brighton: Sussex Centre for Migration Research.
Cooper, J. A. G., and Pile, J. (2014). The Adaptation-Resistance Spectrum: A classification of contemporary adaptation approaches to climate-related coastal change. Ocean & Coastal Management 94: 90–98.
Covello, V., and Sandman, P. M. (2001). Risk communication: Evolution and Revolution. In Wolbarst (ed.), Solutions to an environment in peril. Baltimore, MD: John Hopkins University Press: 164–178.
Cunfer, G. (2005). On the great plains: agriculture and environment. College Station, TX: Texas A&M University Press.
Dang, H. L., Li, E., Bruwer, J., and Nuberg, I. (2014). Farmers' perceptions of climate variability and barriers to adaptation: Lessons learned from an exploratory study in Vietnam. Mitigation and Adaptation Strategies to Global Change 19: 531–548.
Dasgupta, S., Huq, M., and Khan, Z. H. (2011). Climate proofing infrastructure in Bangladesh: The incremental cost of limiting future flood damage. Journal of Environment & Development 20: 167–190.
De Haas, H. (2007). The complex role of migration in shifting rural livelihoods: The case of a Moroccan oasis. In T. Naerssen, v. E. Spaan, and A. Zoomers (eds.), Global migration and development. New York/London: Routledge.
———. (2008). Migration and development: A theoretical perspective, Working papers, Paper 9. Oxford: International Migration Institute.
———. (2009). Mobility and human development. Geneva: UNDP.
———. (2010). The internal dynamics of migration processes: A theoretical inquiry. Journal of Ethnic and Migration Studies 36: 1587–1617.
———. (2012). The migration and development pendulum: A critical view on research and policy. International Migration 50: 8–25.
De Jong, G. F. (2000). Expectations, gender, and norms in migration decision-making. Population Studies 54: 307–319.
De Jong, G. F., and Fawcett, J. T. (1981). Motivations for migration: An assessment and a value-expectancy research model. In G. F. De Jong and R. W. Gardner (eds.), Migration decision making: Multidisciplinary approaches to microlevel studies in developed and developing countries. Pergamon: New York: 13–58.
De Jong, G. F., Root, B. D., and Gardner, R. W. (1986). Migration intentions and behavior: Decision making in a rural Philippine province. Population and Environment 8: 41–62.
De Sherbinin, A., Castro, M., Gemenne, F., Cernea, M. M., Adamo, S., Fearnside, P. M., Krieger, G., Lahmani, S., Oliver-Smith, A., Pankhurst, A., Scudder, T., Singer, B., Tan, Y., Wannier, G., Boncour, P., Ehrhart, C., Hugo, G., Pandey, B., and Shi, G. (2011). Preparing for resettlement associated with climate change. Science 334: 456–457.

De Sherbinin, A., VanWey, L. K., McSweeney, K., Aggarwal, R., Barbierie, A., Henry, S., et al. (2008). Rural household demographics, livelihoods and the environment. Global Environmental Change 18: 38–53.

Department of Disaster Management (2012). Disaster Management Information Center (DMIC) e-library on disaster management, Disaster Management Act 2012. [online] Available at: www.kmp.dmic.org.bd/bitstream/handle/123456789/228/217.%20 DM%20Act%2c%202012.pdf?sequence=1 [Accessed on 20 July 2015.]

Dijksterhuis, A. and van Knippenberg, A. (April 1998). "The relation between perception and behavior, or how to win a game of Trivial Pursuit.". Journal of Personality and Social Psychology 74 (4): 865–877

Disaster Management Bureau (2010) National Plan for Disaster Management 2010-2015, Dhaka: Disaster Management & Relief Division, Government of the People's Republic of Bangladesh

Dodman, D. (2009). Analytical review of the interaction between urban growth trends and environmental changes, Paper 1. Urban density and climate change, Revised draft – April 2. New York: UNFPA.

Dodman, D., and Mitlin, D. (2013). Challenges for community-based adaptation: Discovering the potential for transformation. Journal of International Development 25: 640–659.

Douglas, M., and Wildawsky, A. (1982). Risk and culture: An essay on the selection of technological and environmental dangers. Oakland, CA: University of California Press.

Drabo, A. and Mbaye, L.(2011). Climate Change, Natural Disasters and Migration: An Empirical Analysis in Developing Countries, IZA Discussion Papers 5927, Bonn: Institute for the Study of Labor.

Dun, O., and Gemenne, F. (2008). Defining ‘environmental migration. Forced Migration Review 31: 10–11.

EACH-FOR (2009). EACH-FOR Environmental change and forced migration scenarios, D.3.4 Synthesis Report, Bonn: UNU-EHS.

El-Hinnawi, E. (1985). *Environmental refugees*. Nairobi: United Nations Environmental Program.

Ellickson, R. C. (2001). The market for social norms. *American Law and Economic Review* 3: 1–49.

EM-DAT. (2015a). *Disaster data of Bangladesh, The OFDA/CRED International Disaster Database*. Brussels: Universite´ Catholique de Louvain. [online] Available at: www.emdat.be/database [Accessed on 20 July 2015.]

EM-DAT. (2015b). Top 10 natural disasters in Bangladesh for the period 2002 to 2015, sorted by number of casualties, The OFDA/CRED International Disaster Database. Brussels: Universite´ Catholique de Louvain. [online] Available at: www.emdat.be/result-country-profile#sumtable [Accessed on 15 July 2015.]

Entwisle, B., Rindfuss, R. R., Walsh, S. J., Malanson, G. P., Mucha, P. J., Frizzelle, B. G., McDan-iel, P. M., Yao, X., Williams, N. E., Heumann, B. W., Verdery, B. W., Prasartkul, P., Sawangdee, Y., and Jampaklay, A. (2011). *Extreme climate events and migration: An agent-based modelling approach*, extended abstract for submission to: PAA 2012. [Online] Available at: http://paa2012.princeton.edu/papers/122100. [Accessed on 6 July 2013.]

Entwisle, B., Rindfuss, R. R., Walsh, S. J., and Page, P. H. (2008). Population growth and its spatial distribution as factors in the deforestation of Nang Rong, Thailand. *Geoforum* 39: 879–897.

Entwisle, B., Walsh, S. J., and Rindfuss, R. R. (2005). Population and upland crop production in Nang Rong, Thailand. *Population and Environment* 26: 449–470.

Epstein, S. (1994). Integration of the cognitive and the psychoanalytic unconscious. *American Psychologist* 49: 709–724.
Etzold, B., Ahmed, A.U., Hassan, S.R., and Neelormi, S.(2013) Clouds gather in the sky, but no rain falls. Vulnerability to rainfall variability and food insecurity in
Northern Bangladesh and its effects on migration, *Climate and Development*: 6:
18-27
Fankhauser, S., Smith, J. B., and Tol, R. S. J. (1999). Weathering climate change: Some simple rules to guide adaptation decisions. *Ecological Economics* 30: 67–78.
Farm Security Administration. *Office of war information black-and-white negatives.* Search Results for "migration". [online] Available at: www.loc.gov/pictures/search/?q=migration&co=fsa [Accessed on 22 July 2015.]
Fearon, J. D. (1999). *What is identity (as we now use the word)?* Unpublished manuscript, Stanford University. [Online]. Available at: www.stanford.edu/~jfearon/papers/iden1v2.pdf
Feldman, S., and Geisler, C. (2011). *Land grabbing in Bangladesh: In-situ displacement of peasant holdings.* Paper presented at the International Conference on Global Land Grabbing, Organised by the Land Deals Politics Initiative (LDPI) in collaboration with the Journal of Peasant Studies and hosted by the Future Agricultures Consortium at the Institute of Development Studies, University of Sussex, 6–8 April.
Ferris, E. (2012). *Protection and planned relocations in the context of climate change*, Brookings-LSE Project on Internal Displacement, Legal and Protection Policy Series, Division of International Protection, PPLA/2012/04. Geneva: UNHCR.
Finaince Division (2014) Climate Fiscal Framework, Dhaka: Ministry of Finance
Government of the People's Republic of Bangladesh
Findlay, A. M. (2011) Migrant destinations in an era of environmental change, *Global Environmental Change* 21S: S50–S58
____________(2012) Migration: Flooding and the scale of migration *Nature Climate Change* 2: 401-402Findley, S. E. (1994). Does drought increase migration? A study of migration from rural Mali during the 1983–1985 drought. *International Migration Review*, 28(3), 539–553.
Findlay (2011)_Migrant destinations in an era of environmental change, *Global Environmental Change* 21S: S50–S58
Findlay, A., and Geddes, A. (2011). Critical view on the relationship between climate change and migration: Some insights from the experience of Bangladesh. In E. Piguet, A. Pécoud, and P. Guch-teneire (eds.), *Migration and climate change*. Cambridge: Cambridge University Press: 138–159.
Fishback, P. V., Horrace, W. C., and Kantor, S. (2006). The impact of new deal expenditures on mobility during the great depression. *Explorations in Economic History* 43: 179.
Fletcher, L. E., Pham, P., Stover, E., and Vinck, P. (2007). Latino workers and human rights in the aftermath of Hurricane Katrina. *Berkeley Journal of Employment & Labour Law* 107: 107–167.
Food and Agriculture Organization (2011a). *Irrigation in Southern and Eastern Asia in figures – AQUASTAT survey – 2011*. Rome: FAO. [online] Available at: www.fao.org/nr/water/aquastat/basins/gbm/gbm-CP_eng.pdf [Accessed on 17 July 2015.]
________________(2011b) Ganges-Brahmaputra-Meghna Basin [online] Available at: http://www.fao.org/nr/water/aquastat/basins/gbm/index.stm [Accessed on 10 May 2017.]
Foreign and Commonwealth Office (2014). *Human rights and democracy: The 2013 human rights and democracy report*. London: Foreign and Commonwealth Office.

[online] Available at: www.gov.uk/government/case-studies/country-case-study-bangladesh-political-violence [Accessed on 21 July 2015.]

Foresight (2011). *Migration and global environmental change*, Final project report. London: Government Office for Science.

Frerks, G., Warner, J., and Weijs, B. (2011). The politics of vulnerability and resilience. *Ambiente & Sociedade Campinas* 14: 105–122.

Garai, J. (2016). Gender specific vulnerability in climate change and possible sustainable livelihoods of coastal people: A case from Bangladesh. *Revista de Gestão Costeira Integrada* 16: 79–88.

Garden, I., Lewis, J., and Young, K. (1993). Perspectives on policy analysis. In M. Hill (ed.), *The policy process: A reader*. Hertfordshire: Prentice Hall/Harvester Wheatsheaf: 5–9.

Gardner, K. (2009). Lives in motion: The life-course, movement and migration in Bangladesh. *Journal of South Asian Studies* 4(2): 229–251.

Gayathri R, Bhaskaran, P.K. and Sen,D.(2015) Numerical study on Storm Surge and associated Coastal Inundation for 2009 AILA Cyclone in the head Bay of Bengal. *Aquatic Procedia* 4: 404 – 411

Gemenne, F. (2010). *Migration, a possible adaptation strategy*? Policy Briefs, no. 3. Paris: IDDR.

———. (2011). Why the numbers don't add up: A review of estimates and predictions of people displaced by environmental changes. *Global Environmental Change* 21: S41–S49.

Gerlitz, J. Y., Banerjee, S., Hoermann, B., Hunzai, K., and Macchi, M. (2014). Assessing poverty, vulnerability and adaptive capacity: Development of a system to delineate poverty, vulnerability, and adaptive capacity in the Hindu Kush-Himalayas. ICIMOD: Kathmandu.

Global Humanitarian Forum (2009). *The anatomy of a silent crisis*. New York: GHF.

Goh, A. H. (2012). A literature review of the gender-differentiated impacts of climate change on women's and men's assets and well-being in developing countries (No. 106, p. 38), Capri Working Paper.

Gowda, R., and Fox, J. (eds.) (2002). *Judgments, decisions, and public policy*. Cambridge: Cambridge University Press.

Gray, C., and Mueller, V. (2012). Natural disasters and population mobility in Bangladesh. *Proceedings of the National Academy of Sciences* 109: 6000–6005.

Greenpeace (2008). *Blue alert: Climate migrants in South Asia*. Bangalore: Greenpeace India.

Grothmann, T., and Patt, A. (2003). Adaptive capacity and human cognition,

Prepared for presentation at the Open Meeting of the Global Environmental Change Research Community, Montreal, Canada, 16-18 October, Potsdam: Potsdam Institute for Climate Impact Research

———. (2005). Adaptive capacity and human cognition: The process of individual adaptation to climate change. *Global Environmental Change* 15: 199–213.

Grothmann, T., and Reusswig, F. (2006). People at risk of flooding: Why some residents take action and while others do not. *Natural Hazards* 38: 101–120.

Guchhogram (2010). Guchhogram (CVRP) project: Objective of the project, Government of Bangladesh. [online] Available at: www.guchhogram.gov.bd/index.php?option=com_content&task=view&id=462&Itemid=481 [[Accessed on 20 July 2015.]

Guchhogram CVRP Project (2013a). List of Guchhogram established upto 2012–2013. [online] Available at: www.guchhogram.gov.bd/index.php?option=com_content&task=view&id=477&Itemid=501 [Accessed on 16 January 2014.]

———. (2013b). About Guchhogram. [online] Available at: www.guchhogram.gov.bd/index.php?option=com_content&task=view&id=463&Itemid=482 [Accessed on 16 January 2014.]
Guolo, A. (2008). Robust techniques for measurement error correction: A review. Statistial Methods in Medical Review 17: 555–580
Guzman, J. M., Martine, G., McGranahan, G., Schensul, D., and Tacoli, C. (2009). Population dynamics and climate change. New York: UNFPA.
Habiba, U., Shaw, R., and Takeuchi, Y. (2011). Drought risk reduction through SIP approach in northwestern region of Bangladesh. Environmental Hazards 10: 121–138.
Halliday, T. (2008). Migration, risk and the intra-household allocation of labor in El Salvador. Bonn: Institute for the Study of Labor.
Haque, M. E., and Islam, M. M. (2012). Rural to urban migration and household living conditions in Bangladesh. Dhaka University Journal of Science 60: 253–257.
Harmeling, S. (2012). Global climate risk index 2012: Who suffers the most from extreme weather events? Weather-related loss events in 2010 and 1991 to 2010. Germanwatch. [online] Available at: http://germanwatch.org/en/3667 [Accessed on 12 November 2012.]
Heitzman, J., and Worden, R. (eds.) (1989). Bangladesh: A country study. Washington, DC: GPO for the Library of Congress. [online] Available at http://countrystudies.us/bangladesh/ [Accessed on 27 January 2014.]
Henry, S., and Dos Santos, S. (2012). Rainfall variations and child mortality in the Sahel: Results from a comparative event history analysis in Burkina Faso and Mali. Population and Environment 34: 431–459.
Henry, S., Piché, V., Ouédraogo, D., and Lambin, E. (2004). Descriptive analysis of the individual migratory pathways according to environmental typologies. Population and Environment 25: 397–422.
Henry, S., Shoemaker, B., and Beauchemin, C. (2004). The impact of rainfall on the first out-migration: A multi-level event-history analysis in Burkina Faso. Population and Environment 25: 423–460.
Heremele, K. (1997). The discourse on migration and development. In T. Hammar, G. Brochmann, K. Tamas, and T. Faist (eds.), International migration, immobility and development: Multidisciplinary perspective. Oxford: Berg: 133–158.
Holling, C. S. (1973). Resilience and stability of ecological systems. Annual Review of Ecology and Systematics 4: 1–23.
Homer-Dixon, T., and Percival, V. (1996). Environmental scarcity and violent conflict: Briefing book. Toronto: University of Toronto.
Hugo, G. J. (1996). Environmental concerns and international migration. International Migration Review 30: 105–131.
Hulme, M (2008). Geographical work at the boundaries of climate change. Transactions of the Institute of British Geographers 33: 5–11.
Hunter, L. M. (2005). Migration and environmental hazards. Population and Environment 26: 273–302.
Hunter, L. M., and David, E. (2011). Displacement, climate change and gender. In E. Piguet, A. Pe´coud, and P. Guchteneire (eds.), Migration and climate change. Cambridge, MA: Cambridge University Press: 306–330.
Huq, S. (2001). Climate change and Bangladesh. Science 294: 1617.
Huq, S., and Ayers, J. (2008). Climate change impacts and responses in Bangladesh, Note made for the European Parliament, IP/A/CLIM/NT/2007-09 PE 400.990
Huq, S., Rahman, A., Konate, M., Sokona, Y., and Reid, H. (2003). Mainstreaming adaptation to climate change in least developed countries. London: IIED.

Hutton, D. and Haque, C.E. (2004) 'Human vulnerability, dislocation and resettlement: adaptation processes of river-bank erosion-induced displaces in Bangladesh'. Disasters, 28(1). pp. 41–62.

Hutton, D., and Haque, C. E. (2003). Patterns of coping and adaptation among erosion-induced displacees in Bangladesh: Implications for hazard analysis and mitigation. Natural Hazards 29: 405–421.

Iannantuono, A., and Eyles, J. (1997). Meanings in policy: A textual analysis of Canada's "Achieving health for all" document. Social Science & Medicine 44(11): 1611–1621.

Iaquinta, D. L., and Drescher, A. W. (2000). Defining the peri-urban: Rural-urban linkages and institutional connections. Land Reform, Land Settlement and Cooperatives (FAO) 2: 8–26.

Immerzeel, W. W., Pellicciotti, W. and Bierkens, M. F. P. (2013) Rising river flows throughout the twenty-first century in two Himalayan glacierized watersheds Nature Geoscience 6: 742–745.

International Monetary Fund (2005). IMF country report No. 05/410. Washington, DC: International Monetary Fund [online]. Available at: http://planipolis.iiep.unesco.org/upload/Bangladesh/PRSP/Bangladesh%20PRSP%202005%20Unlocking%20the%20potential.pdf [Accessed on 20 July 2015.]

International Organization for Migration (2003). World Migration Report 2003: Managing Migration - Challenges and Responses for People on the Move, Geneva: IOM

Intergovernmental Panel on Climate Change (1990): Policymakers' summary of the potential impacts of climate change. Report from Working Group II to IPCC.

———. (2007a). Climate change 2007: Working group II: Impacts, adaptation and vulnerability, Glossary A-D. [online]. Available at: www.ipcc.ch/publications_and_data/ar4/wg2/en/annexessglossary-a-d.html [Accessed on 22 July 2015.]

———. (2007b). Climate change 2007, synthesis report: Summary for policymakers. Geneva: Intergovernmental Panel on Climate Change. [online] Available at: www.ipcc.ch/pdf/assessment-report/ar4/syr/ar4_syr_spm.pdf [Accessed on 22 July 2015.]

———. (2012). Managing the risks of extreme events and disasters to advance climate change adaptation. A special report of working groups I and II of the Intergovernmental Panel on Climate Change. Cambridge: Cambridge University Press.

———. (2014). Climate change 2014: Synthesis report. Contribution of working groups I, II and III to the fifth assessment report of the Intergovernmental Panel on Climate Change [Core Writing Team, R. K. Pachauri and L. A. Meyer (eds.)]. Geneva: IPCC.

Islam, A. S. (2013). Analysing changes of temperature over Bangladesh due to global warming using historic data, Bangladesh University of Engineering the Technology [online]. Available at: teach-er.buet.ac.bd/akmsaifulislam/publication/Paper_TWAS_Islam.pdf [Accessed on 19 July 2015]

Islam, R., and Hasan, I. (2016). Climate-induced human displacement: A case study of Cyclone Aila in the south-west coastal region of Bangladesh. Natural Hazards 81: 1051–1071.

Islam, M. S., and Islam, A. M. Z. (1985) A Brief Account on Bank Erosion, Model Studies and Bank Protective Works in Bangladesh. REIS Newsletter 2: 10-13.

Jacobson, J. (1988). Environmental refugees: A yardstick for habitability. Washington, DC: Worldwatch Institute.

Jakobeit, C., and Methmann, C. (2012). Climate refugees' as dawning catastrophe? A critique of the dominant quest for numbers. In J. Scheffran, M. Brzoska, H. G. Brauch, M. Link, and J. Schilling (eds.), Climate change, human security and violent conflict. Hexagon Series on Human and Environmental Security and Peace, Vol. 8: 301–314.

Johnson, V. (1947). Heaven's tableland: The dust bowl story. New York: Farrar, Straus and Company.

Kang, Y. H. (2012). Internal migration and adaptation policy options in coastal Bangladesh, Unpublished master's thesis, School of Global Studies, University of Sussex, Brighton.

Karmalkar, A. C., McSweeney, M. N., and Lizcano, G. (n.d.). Climate change country profiles: Bangladesh. [online] Available at: www.geog.ox.ac.uk/research/climate/projects/undp-cp/UNDP_reports/Bangladesh/ Bangladesh.hires.report.pdf. [Accessed on 8 November 2012.]

Keeley J, Scoones I (1999) Understanding Environmental Policy Processes: A Review. IDS Working Paper No 89, Brighton: Institute of Development Studies (IDS).

Kelman, I. (2015). Difficult decisions: Migration from small island developing states under climate change. Earth's Future 3: 133–142.

Kibreab, G. (1997). Environmental causes and impact of refugee movements: A critique of the current debate. *Disasters* 21: 20–38.

Kinzig, A. P., Ehrlich, P. R., Alston, L. J., Arrow, K, Barrett, S., Buchman, T. G., Daily, G.C, Levin, B., Levin, S., Oppenheimer, M., Ostrom, E., and Saari, S. (2013). Social norms and global environmental challenges: The complex interaction of behaviors, values, and policy. *BioScience* 63: 164–175.

Kniveton, D., Rowhani, P., and Martin, M. (2013). *Future migration in the context of climate change*, Briefing paper No 3. Dhaka: Refugee and Migratory Movements, and Brighton: Sussex Centre for Migration Research, University of Sussex.

Kniveton, D R, Layberry, R, Williams, C.J. R. and Peck, M. (2009) Trends in the start of the wet season over Africa. *International Journal of Climatology*, 29: 1216-1225.

Kniveton, D., Schmidt-Verkerk, K., Smith, C., and Black, B. (2008). Climate *change and migration: Improving methodologies to estimate flows*, No 33, IOM Migration Research Series. Geneva: IOM.

Kniveton, D. R., Smith, C. D., and Black, R. (2012). Emerging migration flows in a changing climate in dryland Africa. *Nature Climate Change* 2: 444–447.

Kniveton, D. R., Smith, C., and Wood, S. (2011). Agent based model simulations of future changes in migration flows for Burkina Faso. *Global Environmental Change* 21: S34–S40.

Kothai, U.(2013). Political discourses of climate change and migration: resettlement policies in the Maldives, *The Geographical Journal* 180: 130-140

Kuhn, R. (2003). Identities in motion: Social exchange networks and rural-urban migration in Bangladesh. Contributions to Indian Sociology, 37, 311.

Kreft, S., Eckstein, D., Junghans, L., Kerestan, C., and Hagen, U. (2015). *Global climate risk Index 2015: Who suffers most from extreme weather events?* Weather-related Loss Events in 2013 and 1994 to 2013. Bonn: Germanwatch

Kreibich, V. (2012). Book review of Siddiqui, K., Ahmed, J., Siddique, K., Huq, S., Hossain, A., Nazimud-Doula, S., and Rezawana, N. (2010). *Social formation in Dhaka, 1985–2005: A longitudinal study of society in a Third World megacity*. Farnham: Ashgate *International Journal of Urban and Regional Research* 36: 1349–1364.

Kuruppu, N. (2009). Adapting water resources to climate change in Kiribati: The importance of cultural values and meanings. *Environmental Science & Policy* 2: 799–809.

Kuruppu, N., and Liverman, D. (2011). Mental preparations for climate change adaptation: The role of cognition and culture in enhancing adaptive capacity of water management in Kiribati. *Global Environmental Change* 21: 657–669.

Kuruppu N, Willie R. (2014) Barriers to reducing climate enhanced disaster risks in least developed country-small islands through anticipatory adaptation. *Weather Clim Extremes* 7: 72–83

Laczko, F., and Aghazarm, C. (eds.) (2009). *Migration, the environment and climate change: Assessing the evidence*, Migration Research Series paper 33, International Organization for Migration, Geneva.

Lazar, A., Clarke, D., Adams, H., Akanda, A. R., Szabo, S., Nicholls, R. J., Matthews, Z., Begum, D., Saleh, A. F. M., Abedin, M. A., Payo, A., Streatfield, P.K., Hutton, C., Mondale, M. S., and Moslehuddin, A. Z. M. (2015). Agricultural livelihoods in coastal Bangladesh under climate and environmental change – a model framework. *Environmental Science Processes & Impacts* 17: 1007–1192.

Leiserowitz, A. (2006). Climate change risk perception and policy preferences: The role of affect, imagery, and values. Climatic Change 77: 45–72.

Levine, N. (2010). Crimestat: A spatial statistics program for the analysis of crime incident locations (v 3.3). Washington, DC: Ned Levine & Associates, Houston, TX, and the National Institute of Justice.

Lonergan, S. (1998). The role of environmental degradation in population displacement. Environmental Change and Security Project Report 4: 5–15.

Lu, X., Bengtsson, L., and Holme, P. (2012). Predictability of population displacement after the 2010 Haiti earthquake. Proceedings of the National Academy of Sciences 109: 11576–11581.

Lutz, A. F., Immerzeel, W. W., Shrestha, A. B., and Bierkens, M. F. P. (2014). Consistent increase in High Asia's runoff due to increasing glacier melt and precipitation. Nature Climate Change 4: 587–592.

Mahapatra, S. K., and Ratha, K. C. (2014). Sovereign states and surging water: Brahmaputra river be-tween China and India, Nota Di Lavaro 46. 2015. [online] Available at: www.feem.it/userfiles/attach/2015520162374NDL2015-046.pdf [Accessed on 25 September 2015.]

Mallick, B., and Vogt, J. (2012). Cyclone, coastal society and migration: Empirical evidence from Bangladesh. International Development Planning Review 34: 217–240.

Mallick, B and Sultana, Z. and (2017) Livelihood after Relocation – Evidences of Guchcha Gram Project in Bangladesh, Working paper, [online] available at https://www.researchgate.net/publication/315656947 Accessed on 13 May 2017

Mallick,B. and Etzold, B (2015) Environment, in Mallick,B.and Etzold, B. (Eds.),Migration and Adaptation – Evidence and Politics of Climate Change in Bangladesh, Dhaka: AHDPH Publishing House: 1–17.

Marshall, R., and Rahman, S. (2013). Internal migration in Bangladesh: Character, drivers and policy issues. Dhaka: UNDP.

Martin, M. (2015). How do you leave a sinking island? Thomson Reuters Foundation website. [online] Available at: www.trust.org/item/20150310112224-syrbf 10 March [Accessed on 17 March 2015.]

Martin, M., Billah, M., Siddiqui, T., Kniveton, D., and Black, R (2014). Climate- related migration in rural Bangladesh: A behavioural model. Population and Environment 36: 85–110.

Martin, M., Kang, Y. H., Billah, M., Siddiqui, T., Black, R., and Kniveton, D. (2013). Policy analysis: Climate change and migration Bangladesh, Working paper 4. Dhaka: Refugee and Migratory Movements, and Brighton: Sussex Centre for Migration Research, University of Sussex.

Martin, M., Kang, Y. H., Billah, M., Siddiqui, T., Black, R., and Kniveton, D. (2017). Climate-influenced migration in Bangladesh: the need for a policy realignment, Development Poliyc Review DOI: 10.1111/dpr.12260

Marchiori, L., Maystadt, J.F., Schumacher, I. (2012) The impact of weather anomalies on migration in sub-Saharan Africa, Journal of Environmental Economics and Management 63:355-374

Massey, D. S. (1992). A place called home. New Formations 17: 3–15.

Massey, D. S., Arango, J., Hugo, G., Kouaouci, A., Pellegrino, A., and Taylor, J. E. (1993). Theories of migration: A review and appraisal. Population and Development Review 19: 431–466.

McCarthy, J., Canziani, O., Leary, N., Dokken, D., and White, K. (eds.) (2001). Climate change 2001: Impacts, adaptation and vulnerability. Contribution of working group III to the Third assessment report of the Intergovernmental Panel on Climate Change. Cambridge: Cambridge University Press.

McGregor, J. (1993). Refugees and the environment. In B. Black and V. Robinson (eds.), Geography and refugees. London: Belhaven Press.

McGregor, JoAnn (1994) Climate change and involuntary migration: implications for food security. Food Policy, 19: 120-132.

McLeman, R. (2011). Settlement abandonment in the context of global environmental change. Global Environmental Change 21: S108–S120.

———. (2014). Climate and human migration: Past experiences, future challenges. New York: Cambridge University Press.

McLeman, R., and Smit, B. (2006). Migration as an adaptation to climate change. Climatic Change 76: 31–53.

McLeman, R., Dupre, J., Ford, L. B., Ford, J., Gajewski, K., and Marchildon, G. (2014). What we learned from the Dust Bowl: Lessons in science, policy, and adaptation. Population and Environment 35: 417–440.

McWilliams, C. (1942). Ill fares the land: Migrants and migratory labor in the United States. Boston, MA: Little, Brown and Company.

Mehedi, H., Nag, A. K., and Farhana, S. (2010). Climate induced displacement: Case study of cyclone Aila in the southwest coastal region of Bangladesh. Khulna: Humanity Watch.

Meyer, D.E.; Schvaneveldt, R.W. (1971). "Facilitation in recognizing pairs of words: Evidence of a dependence between retrieval operations". Journal of Experimental Psychology. 90: 227–234

Miah, M., Khan, M.A.M., Rahman, M.(2014) Recent Trends of International Migration and Remittance Flows: An Empirical Evidence of Bangladesh. IOSR Journal of Economics and Finance 2: 16-23.

Migration between Africa and Europe (2013). MAFE project policy briefing No. 1. Brighton: Sussex Centre for Migration Research, and Paris: Institut National d'étude Démographique.

Milan, A., and Ruano, S. (2014). Rainfall variability, food insecurity and migration in Cabrican, Guatemala. Climate and Development 6: 61–68.

Miller, F., Osbahr, H., Boyd, E, Thomalla, F., Bharwani, S., Ziervogel, G., Walker, B., Birkmann, J., Van der Leeuw, S., Rockström, J., Hinkel, J. Downing, T., Folke, C., and Nelson, D. (2010). Resilience and vulnerability: complementary or conflicting concepts? Ecology and Society 15. [online] Available at: www.ecologyandsociety.org/vol15/iss3/art11/ [Accessed on 16 June 2015]

Ministry of Disaster Management and Relief (2013). Bangladesh Hyogo Framework for Action (HFA) monitoring and review through a multi stakeholder engagement process 2011–2013, Interim Report for the period of January 2011–August 2012.

———. (2015) Comprehensive Disaster Management Programme 2010-2015, Annualm Progress Report, January – December 2014: Dhaka: MDMR, Government of the People's Republic of Bangladesh.

Ministry of Environment and Forests (2005). National adaptation programme of action, August. Dhaka: MOEF, Government of the People's Republic of Bangladesh.

———. (2008). Bangladesh climate change strategy and action plan 2008, MOEF, Government of the People's Republic of Bangladesh.

———. (2009). National adaptation programme of action, Updated Version of 2005, August. Dhaka: MOEF, Government of the People's Republic of Bangladesh.

Ministry of Finance (2014). Ministry of expatriates' welfare & overseas part of the budget 2014–2015 statement. [online] Available at: www.mof.gov.bd/en/budget/14_15/gender_budget/en/16_65_Expatriate's%20Welfare%20and%20Overseas_English.pdf [Accessed on 29 July 2015.]

Ministry of Food and Disaster Management (2008). National disaster management policy. Dhaka: Ministry of Food and Disaster Management, Government of the People's Republic of Bangladesh.

Ministry of Water Resources (2005). Coastal zone policy 2005. Dhaka: MOWR, The People's Republic of Bangladesh.

Minogue, M. (1983). Theory and practice in public policy and administration. Policy & Politics 11: 63–85.

Mortreux, C., and Barnett, J. (2009). Climate change, migration and adaptation in Funafuti, Tuvalu. Global Environmental Change 19: 105–112.

Moses, J. (2009). Leaving poverty behind: A radical proposal for developing Bangladesh through Emigration. Development Policy Review 27: 457–479.

Munshi, K. (2003). Networks in the modern economy: Mexican migrants in the US labour market. Quarterly Journal of Economics 118: 549–599.

Muzzini, E., and Aparicio, G. (2013). Bangladesh: The path to middle-income status from an urban perspective. Washington, DC: World Bank.

Myers, N. (1993). Environmental refugees in a globally warmed world. BioScience 43: 752–761.

———. (2002). Environmental refugees: Our latest understanding. Philosophical Transactions of the Royal Society of Biological Sciences 356: 16.1–16.5.

Myers, N., and Kent, J. (1995). Environmental exodus: An emergent crisis in the global arena. Washington, DC: The Climate Institute.

Narayan, D., Patel, R., Schafft, K., Rademacher, A., and Koch-Schulte, S. (2000). *Voices of the poor: Can anyone hear us*? New York: Oxford University Press.

National Aeronautics and Space Administration (2009). *Cyclone Aila, NASA earth observatory*. [online] Available at: http://earthobservatory.nasa.gov/IOTD/view.php?id=38786 [Accessed on 16 June 2015.]

National Research Council (1999). *Human dimensions of global environmental change: Research pathways for the next decade*. Washington, DC: National Academies Press.

———. (2012). *Disaster Resilience: A National Imperative*. Washington, DC: The National Academies Press

Nicholson, C. T. M. (2014). Climate change and the politics of causal reasoning: The case of climate change and migration. *The Geographical Journal* 180: 151–160.

Obokata, R, Veronis, L., and McLeman, R. (2014). Empirical research on international environmental migration: A systematic review. *Population and Environment* 36: 111–135.

Office for the Coordination of Humanitarian Affairs (2012). *Internal displacement – overview*. [online] Available at: www.unocha.org/what-we-do/advocacy/thematic-campaigns/internal-displacement/overview [Accessed on 20 November 2015.]

Oliver-Smith, A. (1996). Anthropological research on hazards and disasters. *Annual Review of Anthropology* 25: 303–328.

(2006). *Disasters and forced migration in the 21st century*. Perspectives from the Social Sciences. SSRC. [online] Available at: http://understandingkatrina.ssrc.org/Oliver-Smith/ [Accessed on 10 March 2015.]

———. (2009). Climate change and population displacement: Disasters and diasporas in the twenty-first century. In S. A. Crate and M. Nuttall (eds.), *Anthropology and climate change: From encounters to actions*. Walnut Creek: Left Coast Press.
———. (2012). Debating environmental migration: Society, nature and population displacement in climate change. *Journal of International Development* 24: 1058–1070.
Oxfam (2009). *The future is here: Climate change in the Pacific*. Victoria, Auckland: Oxfam Australia, Oxfam New Zealand.
Parry, M. L., Canziani, O. F., Palutikof, J. P., van der Linden, P. J., and Hanson, C. E. (eds.) (2007). *Climate change 2007: Impacts, adaptation and vulnerability*, Contribution of Working Group II to the Fourth assessment report of the Intergovernmental Panel on Climate Change. Cambridge: Cambridge University Press.
Paul S.K. & Routray J.K. 2010a. Flood proneness and coping strategies: the experiences of two villages in Bangladesh. *Disasters*, 34: 489−508.
Paul, S., and Routray, J. (2011). Household response to cyclone and induced surge in coastal Bangladesh: Coping strategies and explanatory variables. *Natural Hazards* 57: 477–499.
Peduzzi, P., Chatenou, B., Dao, H., De Bono, A., Herold, C., Kossin, J., Mouton, F., and Nord-beck, O. (2012). Global trends in tropical cyclone risk. *Nature Climate Change* 2: 289–294.
Penning-Rowsell, E. C., Sultana, S., and Thompson, P. M. (2013). The 'last resort'? Population movement in response to climate-related hazards in Bangladesh. *Environmental Science & Policy* 27s: s44–s59.
Perch-Nielsen, S. L., Bättig, M. B., and Imboden, D. (2008). Exploring the link between climate change and migration. Climatic Change 91: 375–393.
Pervin, M. (2013). Mainstreaming climate change resilience into development planning in Bangladesh, Climate change country report. London: IIED.
Pervin, M., and Moin, M. I. (2014). Synergies in the financial landscape. The Dhaka Tribune, April.
Peterson, T. C., Connolley, W. M., and Fleck, J. (2008). The myth of the 1970s global cooling scientific consensus. Bulletin of American Meteorological Society 89: 1325–1337.
Piguet, E. (2010). Linking climate change, environmental degradation, and migration: A methodological overview. Wiley Interdisciplinary Reviews: Climate Change 1: 517–524.
Planning Commission (2005). Bangladesh: Unlocking the potential, national strategy for accelerated poverty reduction. Dhaka: General Economics Division, Planning Commission Government of People's Republic of Bangladesh.
———. (2008). National strategy for accelerated poverty reduction 2: (FY 2009–2011). Dhaka: Planning Commission, Government of the People's Republic of Bangladesh.
———. (2009). Steps towards change, National Strategy for Accelerated Poverty Reduction 2 (FY 2009–2011). Dhaka: Planning Commission, Government of the People's Republic of Bangladesh.
———. (2010). Outline perspective plan of Bangladesh 2010–2021 (Vision 2021). Dhaka: Planning Commission.
———. (2011). 6th Five Year Plan (2011–2015) Accelerating growth and reducing poverty, Part-1 Strategic Directions and Policy Framework. Dhaka: Planning Commission, Government of the People's Republic of Bangladesh.
———. (2012). The millennium development goals: Bangladesh progress report 2011. Dhaka: Planning Commission, Government of the People's Republic of Bangladesh.
———. (2013). In S. Lockie, D. A. Sonnenfeld, and D. R. Fisher (eds.), Routledge international handbook of social and environmental change. Oxford and New York: Routledge.

Poncelet, A. (2007). Bangladesh case study report: The land of mad rivers, EACH-FOR environmental change and forced migration scenarios. [online] Available at: www.each-for.eu/documents/CSR_bangladesh_090126.pdf. [Accessed on 23 October 2013.]

Poncelet, P., Gemenne, F., Martiniello, M., and Bousetta, H. (2010). A country made for disasters: Environmental vulnerability and forced migration in Bangladesh. In T. Afifi and J. Jäger (eds.), Environment, forced migration and social vulnerability. Berlin | Heidelberg: Springer-Verlag.

PreventionWeb (2012). Introduction to the link to Bangladesh: National plan for disaster management 2010–2015. [online] Available at: www.preventionweb.net/english/professional/policies/v.php?id=16676 [Accessed on February 28, 2014.]

Price, J.C. and Leviston. Z. (2014) Predicting pro-environmental agricultural practices: The social, psychological and contextual influences on land management, Journal of Rural Studies 34: 65 –78

Rahman, A. A., Alam, M., Alam, S. S., Uzzaman, M. R., Rashid, M., and Rabbani, G. (2007). Risks, vulnerability and adaptation in Bangladesh, Occasional paper, Human Development Report 2007/2008 Fighting climate change: Human solidarity in a divided world, Human Development Report Office. [online] Available at: http://climatechange.gov.bd/sites/default/files/Risk%20vulnerability%20and%20adaptation%20in%20Bangladesh_Rahman_et%20al.pdf [Accessed on February 28, 2014.]

Rahman, M. M. (2009). Temporary migration and changing family dynamics: Implications for social development. Population, Space and Place 15: 161–174.

Rahman, M.M., Rafiuddin, M. and Alam, M.M. (2013) Journal of Earth System Science 122: 551.

Raillon, C. (2010). Bangladesh climate disasters, humanitarian practice challenged by populations 'resilience'. Plaisians: Groupe Urgence, Réhabilitation, Développement, and Khulna: Rupantar.

Rajya Sabha (2012). Answer by the Indian Minister of External Affairs S M Krishna to question no 130 in Rajya Sabha, the upper house of the Indian Parliament on March 22, 2012. Available at: www.idsa.in/resources/parliament/RefusalofillegalmigrationbyBangladesh [Accessed on October 4, 2012.]

Rammel, C., Stag, S., and Wolfing, H. (2007). Managing complex adaptive systems – a co-evolutionary perspective on natural resource management. Ecological Economics 63: 9–21.

Ravenstein, E. G. (1885). The laws of migration. Journal of the Royal Statistical Society 48: 167–227.

Reckien, D., Wildenberg, M., and Bachhofer, M. (2013). Subjective realities of climate change: How mental maps of impacts deliver socially sensible adaptation options. Sustainabilty Science 8: 159–172.

Renaud, F., Bogardi, J. J., Dun, O., and Warner, K. (2011). Control, adapt or flee: How to face environmental migration? InterSecTions number 5. Bonn: United Nations University Institute for Environment and Human Security.

Reuveny, R. (2007). Climate change-induced migration and violent conflict. Political Geography 26: 656–673.

———. (2008). Ecomigration and violent conflict: Case studies and public policy implications. Human Ecology 36: 1–13.

Rich, J. T., Neely, J. G., Paniello, R. C., Voelker, C. C. J., Nussenbaum, B., and Wang, E.W (2010). A practical guide to understanding Kaplan-Meier curves. Otolaryngology-Head and Neck Surgery 143: 331–336.

Ritchey, P. N. (1976). Explanations of migration. Annual Review of Sociology 2: 363–404.

Rogaly, B. (2009). Spaces of work and everyday life: Labour geographies and the agency of unorganised temporary migrant workers. Geography Compass 6: 1975–1987.

Rotter, J. B. (1966). Generalized expectancies for internal versus external control of reinforcement. Psychological Monographs 8: 1–28.

Roy, D. (2011). Vulnerability and population displacements due to climate-induced disasters in coastal Bangladesh. In M. Leighton, X. Shen, and K. Warner (eds.), *Climate change and migration: Re-thinking policies for adaptation and disaster risk reduction.* Bonn, Germany: United Nations University Institute for Environment and Human Security (UNU-EHS): 22–31.

Roy, S. S. (2009). A spatial analysis of extreme hourly precipitation patterns in India. *International Journal of Climatology* 29: 345–355.

Salkind, N. J. (2010). *Statistics for people who think they hate statistics*, Second edition, Excel 2007 edition. London: Sage.

, S. O. and Sandberg, K. (2009), Spatial econometric model of natural disaster impacts on human migration in vulnerable regions of Mexico. *Disasters* 33: 591–607

Schmidt-Verkerk, K. (2011). The potential influence of climate change on migratory behaviour – a study of drought, hurricanes and migration in Mexico, Unpublished DPhil Thesis. University of Sussex, Brighton.

Schmuck, H. (2000). "An act of Allah": Religious explanations for floods in Bangladesh as survival strategy. *International Journal of Mass Emergencies and Disasters* 18: 85–95.

Scoones, I. (1998). *Sustainable livelihoods: A framework for analysis*, IDS working paper 72. Brighton: Institute of Development Studies.

Sen, A. (1981). Poverty and famines. An essay on entitlement and deprivation. Oxford: Clarendon Press.

Shahid, S., Harun, S. B., and Katimon, A. (2012). Changes in diurnal temperature range in Bangladesh during the time period 1961–2008. *Atmospheric Research* 118: 260–270.

Shamsuddoha, M., and Chowdhury, R. K. (2007). *Climate change impact and disaster vulnerabilities in the coastal areas of Bangladesh.* Dhaka: COAST Trust, Equity, and Justice Working Group.

de Sherbinin, A. (2014) Climate Change Hotspots Mapping: What Have We Learned? *Climatic Change*, 123, 23-37.

Shumway, M., Otterstroma, S., and Glavac, S. (2014). Environmental hazards as disamenities: Selective migration and income change in the United States from 2000–2010. *Annals of the Association of American Geographers* 104: 280–291.

Siddiqui, T. (2009). *Climate change and population movement: The Bangladesh case.* Paper presented at the Conference on climate insecurities, human security and social resilience, The RSIS Centre for Non-traditional Security Studies, Singapore, 27–28 August.

———. (2011). Rationale of framing new emigration law, Law Update. *The Daily Star*, Dhaka, June 25.

———. (2014). Where is the Migration Law 2013? *The Daily Star*, Dhaka, December 2. [online] Available at: www.thedailystar.net/where-is-the-migration-law-2013-52917 [Accessed on 19 August 2015.]

Siddiqui, T., and Farah, M. (2011). *Facing the challenges of labour migration from Bangladesh.* Protifolon Policy Brief, 4. Dhaka: Institute of Informatics and Development.

Siddiqui, T., Islam, M.T., Akhter, Z (2015) National Strategy on the Management of Disaster and Climate Induced Displacement. Dhaka: Comprehensive Disaster Management Programme II, Ministry of Disaster Management and Relief.

Silverman, B. W. (1986). *Density estimation for statistics and data analysis*, Monographs on Statistics and Applied Probability, School of Mathematics, University of Bath, UK. [online] Available at: http://ned.ipac.caltech.edu/level5/March02/Silverman/Silver_contents.html [Accessed on 19 Aug 2015.]
Simon, H. A. (1996). *The sciences of the artificial*, third edition. Cambridge, MA: MIT Press.
Skeldon, R. (1997). Migration and development: A global interpretation. London: Longman.
Slovic, P. (1987). "Perception of Risk." Science 236(17 April): 280-285
_________(2000). The Perception of Risk: London: Earthscan
Smit, B., and Wendel, J. (2006). Adaptation, adaptive capacity and vulnerability. *Global Environmental Change* 16: 282–292.
Smith, C., Wood, S., and Kniveton, D. (2010). *Agent based modelling of migration decision-making*. Proceedings of the European Workshop on Multi-Agent Systems (EUMAS-2010).
Spaan, E., van Naerssen T. and Hillmann F. (2005) 'Shifts in the European discourses on migration and development', Asian and Pacific Migration Journal, 14: 35-69.
Stal, M., and Warner, K. (2009). *The way forward: Researching the environment and migration nexus*, Research brief based on the outcomes of the 2nd expert workshop on climate change, environment, and migration, 23–24 July 2009, Munich, Germany. UNU-EHS. ISSN: 1816–5788.
Stark, O. (1984). Migration decision making: A review article. *Journal of Development Economics* 14: 251–259.
Stark, O., and Bloom, D. E. (1985). The new economics of labor migration. *American Economic Review* 75: 173–178.
Stark, O., and Levhari, D. (1982). On migration and risk in LDCs. *Economic Development and Cultural Change* 31: 191–196.
Steele, F. (2004). *Event history analysis*, NCRM Methods Review Papers, NCRM/004. Southampton: ESRC National Centre for Research Methods.
Steinbeck, J. (1992). *Grapes of wrath*, Original. New York: Penguin, original work published in 1939.
Stern, N. (2007). *The economics of climate change: The Stern review*. Cambridge: Cambridge University Press.
Suhrke, A. (1993). *Pressure points: Environmental degradation, migration and conflict*. Occasional paper of project on environmental change and acute conflict. Washington, DC: American Academy of Arts and Sciences.
SUHRKE, A. (1994). Environmental degradation and population flows. *Journal of International Affairs*, 47: 474–96.
Sultana, Z. and Mallick, B. (2015) Adaptation Strategies after Cyclone in Southwest Coastal Bangladesh – Pro Poor Policy Choices *American Journal of Rural Development* 3: 24-33
Swain, A. (1996). Displacing the conflict: Environmental destruction in Bangladesh and ethnic conflict in India. *Journal of Peace Research* 33: 189–204.
Sward, J., and Codjoe, S. (2012). *Human mobility and climate change adaptation policy: A review of migration in National Adaptation Programmes of Action (NAPAs)*, Migrating out of poverty research programme consortium, Working paper 6. Brighton: Migrating out of Poverty.
Tacoli, C. (ed.) (2006). *The Earthscan reader in rural urban linkages*. London: Earthscan.
———. (2009). Crisis or adaptation? Migration and climate change in a context of high mobility. *Environment and Urbanization* 21: 513–525.

———. (2011). Not only climate change: Mobility, vulnerability and socio-economic transformations in environmentally fragile areas in Bolivia, Senegal and Tanzania. London: International Institute for Environmental and Development (IIED).

Tan, V (2015) Some 25,000 risk sea crossings in Bay of Bengal over first quarter, almost double from year earlier, News, UNHCR. [online] Available at http://www.unhcr.org/news/latest/2015/5/554c9fae9/25000-risk-sea-crossings-bay-bengal-first-quarter-double-year-earlier.html Accessed on 15 May, 2017.

Tearfund International (2006). *Feeling the heat*. London: Tearfund.

Tickell, C. (1989). *The human impact of global climate change*, Natural Environment Research Council annual lecture at the Royal Society, London, 5 June.

Todaro, M. P. (1980). Internal migration in developing countries: A survey. In R. A. Easterlin (ed.), *Population and economic change in developing countries*. Chicago: University of Chicago Press: 361–402.

———. (2014). World urbanization prospects, the 2014 revision, highlights, economic and social affairs. New York: UN.

UNDP. (2009). Human development report 2009. Overcoming barriers: Human mobility and development. New York: UNDP.

———. (2010). Mobility and migration, a guidance note for Human Development Report teams, Human Development Report Office. New York: UNDP.

UNEP (2003). *Discussion document for the expert think tank meeting, Intergovernmental Panel on Global Environmental Change*. Losby Gods, Oslo, Norway, 15 January 2003. [online] Available at: www.unep.org/scienceinitiative/IPEC_Discussion_Doc.doc [Accessed on 22 July 2015.]

———. (2008). *Natural disasters contribute to rise in population displacement*. [online] Available at: www.unep.org/Documents.Multilingual/Default.asp?DocumentID=538&ArticleID=5842&l=en [Accessed on 21 July 2015.]

UNFCCC (2007). *UNFCCC executive secretary says significant funds needed to adapt to climate change impacts*. [online] Available at: http://unfccc.int/files/press/news_room/press_releases_and_advisories/application/pdf/070406_pressrel_english.pdf [Accessed on 21 July 2015.]

———. (2011). *Report of the conference of the parties on its sixteenth session*, held in Cancun from 29 November to 10 December 2010, Addendum, part two: Action taken by the Conference of the Parties at its sixteenth session.

———. (2014). *Bangladesh experiences with the NAPA process*. [online] Available at: http://unfccc.int/adaptation/knowledge_resources/ldc_portal/bpll/items/6497.php [Accessed on 13 August 2014.]

UNHCR (1993). *The state of the world's refugees 1993: The challenge of protection*. [online] Available at: www.unhcr.org/3eeedcf7a.html [Accessed on 27 November 2015]

———. (2014). *Southeast Asia irregular maritime movements*. January–March 2015. [online] Available at: www.unhcr.org/554c6a746.html [Accessed on 27 July, 2015.]

UNU-EHS (2005). *As ranks of environmental refugees swell worldwide, calls grow for better definition*, recognition, support. [online] Available at: http://archive.unu.edu/media/archives/2005/mre29-05.doc [Accessed on 21 July 2014.]

Van der Geest, K. (2011). *Migration, environment and development in Ghana*. Paper presented at the international conference, Rethinking Migration: Climate, Resource Conflicts and Migration in Europe, 13–14 October. [online] Available at: www.network-migration.org/rethinking-migration-2011/2/papers/Geest.pdf [Accessed on 21 July 2015.]

Walsham, M. (2010). *Assessing the evidence: Environment, climate change and migration in Bangladesh*. Dhaka: International Organization for Migration, Regional Office for South Asia

Warner, K., and Afifi, T. (2014). Where the rain falls: Evidence from 8 countries on how vulnerable households use migration to manage the risk of rainfall variability and food insecurity. *Climate and Development* 6: 1–17.

Warner, K., Afifi, T., Henry, K., Rawe, T., Smith, C., and De Sherbinin, A. (2013). *Where the rain falls: Climate change, food and livelihood security and migration*. An 8-country study to understand rainfall. Bonn: UNU-EHS.

Warner, K., Ehrhart, C., Sherbinin, A., Adamo, S., and Chai-Onn, T. (2009). *In search of shelter: Mapping the effects of climate change on human migration and displacement*. Geneva: CARE International.

Warner, K., Hamza, M., Oliver-Smith, A., Renaud, F., and Julca, A. (2010). Climate change, environmental degradation and migration. *Natural Hazards* 55: 689–715.

Warner, B. P., Kuzdas, C., Yglesias, M. G., & Childers, D. L. (2015). Limits to adaptation to interacting global change risks among smallholder rice farmers in Northwest Costa Rica. *Global Environmental Change* 30: 101–112.

Water Resources and Planning Organisation (2006). *Coastal development strategy*. Dhaka: WARPO, Government of the People's Republic of Bangladesh.

Webster, P. J. et al. (2005). Changes in tropical cyclone number, duration, and intensity in a warming environment. *Science* 309: 1844–1846.

Webster, P.J., J. Jian, T. M. Hopson, C. D. Hoyos, P. Agudelo, H-R. Chang, J. A. Curry, R. L. Grossman, T. N. Palmer, A. R. Subbiah 2010: Extended-range probabilistic forecasts of Ganges and Brahmaputra floods in Bangladesh. *Bulletin of the American Meteorologcial Society*. 91: 1493-1514

Williams, R. (2015). *Interaction effects and group comparisons*, University of Notre Dame. [online] Available at: http://www3.nd.edu/~rwilliam/ [Accessed on 25 April 2016.]

Wisner, B., Blaikie, P., Cannon, T., and Davis, I. (2004). *At risk: Natural hazards, people's vulnerability and disasters*, 2nd edition. New York: Routledge.

Wolpert, J. (1965). Behavioral aspects of the decision to migrate. *Papers and Proceedings of the Regional Science Association* 15: 159–169.

——. (1966). Migration as an adjustment to environmental stress. *Journal of Social Issues* 22(4): 92–102.

World Bank (2010a). *Economics of adaptation to climate change*. Washington, DC: World Bank.

—— (2010b). *World development report 2010: Development and climate change*. Washington, DC: The World Bank.

———. (2012). *World development report 2013: Jobs*. Washington, DC: World Bank.

———. (2015). *Data, by country, Bangladesh*. [online] Available at: http://data.worldbank.org/country/bangladesh. [Accessed on 16 June 2015].

———. (2016). Population density (people per sq. km of land area), Food and Agriculture Organization and World Bank population estimates [online] Available at: *http://data.worldbank.org/indicator/EN.POP.DNST* [Accessed on 14 May 2017].

Worster, D. (1979). *Dust bowl: The southern plains in the 1930s*. New York: Oxford University Press.

———. (1986). The dirty thirties: A study in agricultural capitalism. *Great Plains Q* 6: 107–116.

Yanow, D. (1992). Silences in public policy discourse: Organizational and policy myths. *Journal of Public Administration Research and Theory* 2: 399–423.

Young Power in Social Action (2013). *Bangladesh housing land and property rights initiative*, Report prepared for Displacement Solutions. Chittagong: YPSA.

Zaman, M. Q. (1989). The social and political context of adjustment to riverbank erosion hazard and population resettlement in Bangladesh. *Human Organization* 48: 196–205.

Index

Abrar, C. R. 21, 130
Abu, M. 43
adaptations: definition of 134; international migration as 18; migration as form of 33, 133–4; migration strategies 104–6; motivation as form of 40–2; movement as way of 33; non-movement 42; options 42; policy formulation 47; public policy and 104–6; resilience and 40–2, 114; as socio-cognitive-making process 39–40
adaptive capacity 36–7, 47, 70
Afifi, T. 19
age, as key socio-economic variable 80, 85, 95, 122
Agent Based Model (ABM) 61
agriculture, shift away from 67 *See also* farming
Ajzen, I. 36
assets, as key socio-economic variable 80–1, 95, 100, 127–8
Awami League 115
Azad, S. N. 21, 130

Bangladesh: agriculture in GDP 17; as climate change hotspot 5; climatic and environmental hazards of 14–16; Coastal Zone Policy 110; Comprehensive Disaster Management Programme (CDMP) 110; cyclones and storm surges impacting 15–16; disaster-proofing vulnerable areas of 20; disaster response measures 112–13; disaster vulnerabilities 110–13; economy 66–7; floodplains 14; geographic and demographic features of 13–14; internal migration rate 16; international migration rate 16; liberation struggle of 109; map of *9*, *55*; migration policies 113–15; national GDP of 66; Overseas Employment and Migration Act 2013 113–14; poverty reduction policies 106–10; river flow patterns in 14–15; Sixth Five-Year Plan 108–9; Vision 2021 108–9
Bangladesh Bureau of Statistics 17
Bangladesh Climate Change Resilience Fund (BCCRF) 114–15
Bangladesh Climate Change Strategy and Action Plan (BCCSAP) 114
Barnett, J. 42
barriers to movement 59
Barrios, S. 43
behavioural factors: for better livelihoods and income 66–70; expectations and perceived resources 70–1; perceived level of risk-taking 71–3; understanding migration decisions 65
Bhagyakul village 54, 64, 69
Bhola island 20
Black, R. 28, 39, 44, 59, 61, 132
Blackburn, Simon 31–2
Blaikie, P. 26
Bohra-Mishra, P. 129
bounded rationality 33
Brahmaputra river 14
Brown, L. A. 33
Bureau of Manpower Employment and Training (BMET) 118
Burkina Faso 43, 81
business-as-usual economic activity 72–3, 76, 77

Castles, S. 29
causation models 46
Centre for Research on the Epidemiology of Disasters (CRED) 45–6
Centre for Urban Studies 92
Charipara village 56, 64, 69
Chittagong 17, 99

Chittagong Hill Tract (CHT) 21–2
Chorpka village 54, 67–8, 125
Chor Sonakanda 56
circular migration 20–1
climate- and environment-related migration: in Bangladesh 19; definition of 5; differential impacts of 128–30; fears of 27; Foresight report 6–7; policy implications 46–8; rhetoric of 28–9. *See also* migration
climate change: adapting to 40–2; definition of 4; impacting mobility 4; increasing displacement of people 3; linking to migration 3–4; media coverage of 23; as political tool 41
climate-change-induced conflict 46
Climate Change Trust Fund (CCTF) 114
climate-environment-migration nexus 43–6
climate fiscal framework (CFF) 108
climate-induced migration. *See* climate- and environment-related migration
climate refugees 2, 27, *28*, 61
climate risks, perceptions of 59
climate variability: in Bangladesh 13; definition of 4; limiting adaptation choices 129, 132; migration decisions and 39, 59, 75; reducing agricultural productivity 21; women's livelihood activities and 125
climatic and environmental factors that drive migration *86*, *87*, 123–6
climatic and environmental stresses and shocks: in Bangladesh villages 53; diversifying livelihoods 122; gender roles and 79; increasing short-term rural-to-rural migration 81; influencing human movement 130; large scale displacement and 44; mobility outcomes of 43–6; policy implications 46–8; reducing vulnerabilities 75; risk perceptions and 61
Coastal Zone Policy 110
cognitive analysis of migration decision-making 35–8
Collins, A. E. 132
Collyer, M. 39, 59
Comprehensive Disaster Management Programme (CDMP) 110–11
Cooper, J. A. G. 42
Coordinated Border Management Plan 117
cross-border migration 117
cultural practices, negative role 59
culture-specific norms 73–4
cumulative causation 39
Cunfer, G. 26
Cyclone Aila 1, 15, 52, 53, 56, 64, 69, 124, 129, 132
cyclones: after-effects of 15–16, 64; discouraging movement within district 131; at Gabura 52; Kaplan-Meier survival graph *93*; as major driver of migration 20, 129–30; moving house and 95–9, 101; at Satkhira 53, 56; vulnerability to 30, 35, 52
Cyclone Sidr 15, 53, 56

Dakerhati 56, 67, 68
Dang, H. L. 129
data gathering 8–9
David, E. 73
decision-making. *See* migration decisions/decision-making
De Jong, G. F. 36, 39
demographic pressures for migration 21
deprivation 81
Dhaka *9*, 17, 99
Disaster Management Act 112, 113
disasters: community-level adaptations to 111; definition of 34; management framework 111–13; migration patterns and 41; people affected by 45–6; prompting short-distance moves 128; reconstruction after 132; risk reduction 15, 40, 103–7, 110–13, 131; vulnerabilities 110–13, 124–5
displacement: adaptation interconnected with 134; after environmental shocks 20, 124–5; erosion leading to 64; flooding and 128; as forced migration 108; maladaptation by 40–1; management framework 112; risk of 3
diversifying livelihoods 75, 122, 125
Drabo, A. 44
droughts 53, 64–5, 68, *94*, 128, 131
Dun, O. 29
Durlogpur 54
Dust Bowl phenomenon 25–6

economy, promoting over-exploitation of nature 26
education, negative impact on migration 122–3
El Niño-Southern Oscillation (ENSO) 15
El Salvador 44
Environmental Change and Forced Migration Scenarios (EACH-FOR) 19, 44–5, 129
environmental degradation 28
environmental exodus 27

environmental migration. *See* climate- and environment-related migration
environmental refugees 27, 29
environment-climate-migration nexus: Dust Bowl phenomenon 25–6; early references of 25–7
event history model 78, 79

families and women, migration of 124–5
family size, as key socio-economic variable 80
farming: Bangladesh GDP 66; climatic stress on 18, 44; Dust Bowl phenomenon 25; over- 26; practices 116; rainfall variability 63; salinisation and 20–1, 64, 65, 124, 129–30; shift away from 67–8
Farm Security Administration 25
Findley, A. 43
first migration 83, 85–8, *89*, 95–9, *96*, *97*, *98*
flooding/floods 53, 64, 100, 128–9, 131
flood plain dwellers 35–7
floodplains 14
focus group discussions 57
food scarcity 20–1
food security 19, 44, 79, 115, 128–9, 132–3
forced migration 7, 23–4, 28, 34, 108, 116
Foresight report 6, *6*, 19–20, 45, 127
free choice in migration 32
functionalist theories of migration 30

Gabura village 1, 51–2, 57, 64, 74, 124
Ganges-Brahmaputra-Meghna system 14
Ganges river 14
Gardner, K. 132
Gar Kumarpur village 56
garment factories 2, 67, 69, 90, 118
Gemenne, F. 27, 29
gender-disaggregated models 95
gender norms 73
gender roles 36, 39, 79–80, *90*, 100
Ghana, outmigration in 43
global environmental change 4
global warming 15
Grapes of Wrath (Steinbeck) 25
Gray, C. 18, 127
Grothmann, T. 36, 38–40, 59
Guchhogram Climate Victims Rehabilitation Project 21, 109

Halliday, T. 44
Hatia island 20
hazards 14–16, *87*, 110, 126, 131 *See also* cyclones; flooding/floods; riverbank erosion
Henry, S. 43, 81
heuristics 38–9, 59
Himalayan rivers 14–15
historical-structuralism 31–2
households 75, 127
Hulme, M. 38, 59
Hunter, L. M. 73
Hurricane Katrina 46
Hurricane Mitch 34

identity, definition of 58
immobility: definition of 2; Foresight report 45; in hazard-prone places 126–7, 130–3; MARC model 62; of the poor 127; as serious risk 20
in-depth interviews 57
individual decision-making process 60–2
in situ adaptation 42
intentions 39 *See also* migration decisions/ decision-making
Intergovernmental Panel on Climate Change (IPCC) 3
internal migration: as adaptation strategy 18; adaptive roles of 103, 104; government recognition of 116, 118; households dependent on 67–8; moving house and 101; policy development and 107–9; rate of 16
International Centre for Migration Policy Development (ICMPD) 103–4
internationally planned resettlements 110
international migration 17–18, 24, 117
international migration rate 16
international remittances 17–18

Jakobeit, C. 27
Jamuna river 14
Jessore 99

Kaplan-Meier survival analysis 92–5, *93*, *94*
Kelman, I. 41
Kent, J. 27
Kernel density estimations 88–92, *89*, *90*, *91*
Khesba 54
Khulna *9*, 99, 125
Khutikata 56
Kiribati 59
Kniveton, D. R. 7, 36–7, 44, 61, 127
Kolma 55, 64
Kothari, U. 41

Kuhn, R. 67
Kuruppu, N. 38–9, 59

land degradation 43
land-grabbing 21–2
land tenue 21
livelihood activities 74, 76
Liverman, D. 38–9, 59
locus of control 70–1
logistic regression models 39
logit models: climatic and environmental variables *86*; data quality control 81–3; descriptive statistics 83–4; event history analysis 79; first migration outside the district 85–8; individuals at risk of migration formula 78–9; odds ratio 79; socio-economic variables *86*; using Kaplan-Meir graphs 92–5; using Kernel density estimate graphs 88–92; variables selection 79–81
Lohajong 54–6, *55*
Lonergan, S. 28

maladaptation 40–1
Mandra 55–6, 67
Marchiori, L. 44
mass migrations 26, 29, 46, 48, 134 *See also* migration
maximalists 29
Mbaye, L. 44
McGregor, J. 28
McLeman, R. 134
measurement error correction techniques 82
Meghna Estuary 14–15
Methmann, C. 27
migrant labour 107
migration: as adaptation strategy 104–6; attitudes 66–70; behavioural response to 30–5; benefits of 47; climatic and environmental variables influencing *87*; controls 116; as coping strategy 19; definition of 83; distance and duration of 69; drivers 18–22; free choice in 32; intentions 39; modernisation process of 30–2; multi-causal nature of 19; outside the district *84*, 85; public spending to reduce 26; reasons for 1–2, 7; as socially acceptable behaviour 123. *See also* climate- and environment-related migration
Migration and global environment change: future challenges and opportunities (report) 4
migration control beliefs 70–1
migration decisions/decision-making: behavioural factors 65; cognitive analysis 35–8; consultations with family members 72; in-depth interviews 57; described 7; environment and climate role in 32–5; focus group discussions 57; heuristics and 38–9; individual-level decision-making 58, 60–2; perceptions and expectations of 38–40; process of 59–60; in socio-economic contexts 66; transitioning from rural farming-based livelihoods 58; village surveys 57
migration patterns: across districts 67–8; altering in extreme weather conditions 105; in Chorpka 67–8; general 16–18; male prevalence in 74; rural-to-urban 16; uncertanties of 24
migration personal norms 71–3
migratory pathways 43
Milan, A. 133
Millennium Development Goals (MDG) 109
Ministry of Environment and Forests (MoEF) 114–15
Ministry of Land Administration and Land Reforms 109–10
mobility 4, 43–6, 135
Model of Migration Adaptation to Rainfall Change (MARC) 36–7, *38*, 61–2, 129
Model of Private Proactive Adaptation to Climate Change (MPPACC) 36, *37*, 61
Mohnohil 53–4
Moinul Islam Moin 103
monga 20–1, 128, 133
monsoons 54
Moore, E. G. 33
motivations 18–22, 33–4, 58, 76, 134
movement, barriers to 33, 59
moving house *98*, 100–1 *See also* first migration
Mueller, V. 18, 127
multi-causal nature of migration 19, 25–6, 124, 127–8, 134
multi-donor Trust Fund (MDTF) 114
Munshi, K. 43
Munshiganj *9*, 54, *55*, 56, 64, 67, 124
Mushiganj 53
Myers, N. 27

Nachole 53–4
National Adaptation Programmes of Action (NAPAs) 47, 114–15
National Disaster Management Policy 112
National Plan for Disaster Management 112–13

National Strategy for Accelerated Poverty Reduction 106
National Strategy on the Management of Disaster and Climate Induced Displacement 103, 111–12
Nawabganj *9*, 53, *55*, 64
neoclassical migration theories 30–1
New Economics of Labour Migration (NELM) theory 32
non-movement. *See* immobility

Oliver-Smith, A. 26, 34
Outline Perspective Plan 116
outmigration 43
Overseas Employment and Migration Act 2013 113–14
Overseas Employment Policy 113, 117, 118

Patt, A. 36, 38–40, 59
Paul, S. K. 127, 131
Penning-Rowsell, E. C. 127, 128, 130, 132, 133
peri-urban spaces 107
Pile, J. 42
planned behaviour theory 36
policy realignment 119
poverty 95, 100, 122
Poverty-Environment-Climate Mainstreaming (PECM) 108
Poverty Reduction Plan 107–8
poverty reduction policies 106–10
primary data 8
priming 66
public policy: adaptive strategies 104–6; poverty reduction 106–10

qualitative field data 8
quantitative data 8

Rahman, Mujibur 109–10
rainfall trends 63
rainfall variability 45, 81, 129
Rajshahi 99
Raleigh, C. 46
Rautray, J. K. 127, 131
Ravenstein, E. G. 25
reasoned action theory 36
Refugee and Migratory Movements Research Unit (RMMRU) 8
remittances 17–18, 47, 107
resettlement 113
resilience 40–2
resilience fund 114
Reuveny, R. 46
risk reduction 15, 40, 103–7, 110–13, 131
riverbank erosion: in Charipara 56; at Chorpka village 125; displacement and 20, 56; in Gabura 52; Kaplan-Meier survival graph *94*, 95; loss of homesteads and farms 54; in Mushiganj 53, 64; negative impact on inter-district migration 131; positive influence on short-distance migration 130
rootedness, sense of 132–3, 135–6 *See also* immobility
Ruano, S. 133
rural-to-rural migration 16
rural-to-urban migration 16
rural-urban transition zones 107

Saldaña-Zorrilla, S. O. 44
salinisation 20, 21, 65, 125, 129–30
Sandberg, K. 44
Satkhira *9*, 53, *55*, 69
seasonal deprivation 16, 18, 81
seasonal migration 18, 20, 64, 92, 124, 128
secondary data 8
Sendai Framework of DRR 103
Shibganj 54
shocks. *See* stresses and shocks
Shumway, M. 41
Simon, H. A. 33
Sirajdikhan 56
Sixth Five-Year Plan 108–9, 116–17
slum dwellers 109
small island developing states (SIDS) 41
Smit, B. 134
Smith, Christopher 61
Smith, H. A. 36–7
Social Development Framework (SDF) 112
social networks 74, 80, 95, 101, 122
social norm, definition of 73
social norm beliefs 73–5
socio-economic factors that drive migration 26, 79–81, *86*, 119, 122–3
socio-political reasons for migration 21
spatial equilibrium 41
Sreenagar 54, *55*
Steinbeck, John 25
stresses and shocks. *See* climatic and environmental stresses and shocks
stress-tolerance model of migration 33
structuralism 31–2
Sub-Saharan Africa 44
Sussex Centre for Migration Research 8

trapped populations 45

unattractive climate 25
UNHCR 27

value-expectancy model 33–4
Van der Geest, K. 43
variance-covariance matrix estimation (VCE) 82–3
village surveys 57
village-to-village migration 118
Vision 2021 108–9

Warner, K. 19, 129, 132
Webber, M. 42
Webster, P. J. 15
Wolpert, J. 33
World Bank 20

Zaman, M. Q. 20